Graph Theory with Algorithms and its Applications

Santanu Saha Ray

Graph Theory
with Algorithms
and its Applications

In Applied Science and Technology

 Springer

Santanu Saha Ray
Department of Mathematics
National Institute of Technology
Rourkela, Orissa
India

ISBN 978-81-322-0749-8   ISBN 978-81-322-0750-4   (eBook)
DOI 10.1007/978-81-322-0750-4
Springer New Delhi Heidelberg New York Dordrecht London

Library of Congress Control Number: 2012943969

Printed on acid-free paper

Springer is part of Springer Science+Business Media (www.springer.com)

This work is dedicated to my grandfather late Sri Chandra Kumar Saha Ray, my father late Sri Santosh Kumar Saha Ray, my beloved wife Lopamudra and my son Sayantan

Preface

Graph Theory has become an important discipline in its own right because of its applications to Computer Science, Communication Networks, and Combinatorial optimization through the design of efficient algorithms. It has seen increasing interactions with other areas of Mathematics. Although this book can ably serve as a reference for many of the most important topics in Graph Theory, it even precisely fulfills the promise of being an effective textbook. The main attention lies to serve the students of Computer Science, Applied Mathematics, and Operations Research ensuring fulfillment of their necessity for Algorithms. In the selection and presentation of material, it has been attempted to accommodate elementary concepts on essential basis so as to offer guidance to those new to the field. Moreover, due to its emphasis on both proofs of theorems and applications, the subject should be absorbed followed by gaining an impression of the depth and methods of the subject. This book is a comprehensive text on Graph Theory and the subject matter is presented in an organized and systematic manner. This book has been balanced between theories and applications. This book has been organized in such a way that topics appear in perfect order, so that it is comfortable for students to understand the subject thoroughly. The theories have been described in simple and clear Mathematical language. This book is complete in all respects. It will give a perfect beginning to the topic, perfect understanding of the subject, and proper presentation of the solutions. The underlying characteristics of this book are that the concepts have been presented in simple terms and the solution procedures have been explained in details.

This book has 10 chapters. Each chapter consists of compact but thorough fundamental discussion of the theories, principles, and methods followed by applications through illustrative examples.

All the theories and algorithms presented in this book are illustrated by numerous worked out examples. This book draws a balance between theory and application.

Chapter 1 presents an Introduction to Graphs. Chapter 1 describes essential and elementary definitions on isomorphism, complete graphs, bipartite graphs, and regular graphs.

Chapter 2 introduces different types of subgraphs and supergraphs. This chapter includes operations on graphs. Chapter 2 also presents fundamental definitions of walks, trails, paths, cycles, and connected or disconnected graphs. Some essential theorems are discussed in this chapter.

Chapter 3 contains detailed discussion on Euler and Hamiltonian graphs. Many important theorems concerning these two graphs have been presented in this chapter. It also includes elementary ideas about complement and self-complementary graphs.

Chapter 4 deals with trees, binary trees, and spanning trees. This chapter explores thorough discussion of the Fundamental Circuits and Fundamental Cut Sets.

Chapter 5 involves in presenting various important algorithms which are useful in mathematics and computer science. Many are particularly interested on good algorithms for shortest path problems and minimal spanning trees. To get rid of lack of good algorithms, the emphasis is laid on detailed description of algorithms with its applications through examples which yield the biggest chapter in this book.

The mathematical prerequisite for Chapter 6 involves a first grounding in linear algebra is assumed. The matrices incidence, adjacency, and circuit have many applications in applied science and engineering.

Chapter 7 is particularly important for the discussion of cut set, cut vertices, and connectivity of graphs.

Chapter 8 describes the coloring of graphs and the related theorems.

Chapter 9 focuses specially to emphasize the ideas of planar graphs and the concerned theorems. The most important feature of this chapter includes the proof of Kuratowski's theorem by Thomassen's approach. This chapter also includes the detailed discussion of coloring of planar graphs. The Heawood's Five color theorem as well as in particular Four color theorem are very much essential for the concept of map coloring which are included in this chapter elegantly.

Finally, Chapter 10 contains fundamental definitions and theorems on networks flows. This chapter explores in depth the Ford–Fulkerson algorithms with necessary modification by Edmonds–Karp and also presents the application of maximal flows which includes Maximum Bipartite Matching.

Bibliography provided at the end of this book serves as helpful sources for further study and research by interested readers.

Acknowledgments

I take this opportunity to express my sincere gratitude to Dr. R. K. Bera, former Professor and Head, Department of Science, National Institute of Technical Teacher's Training and Research, Kolkata and Dr. K. S. Chaudhuri, Professor, Department of Mathematics, Jadavpur University, for their encouragement in the preparation of this book. I acknowledge with thanks the valuable suggestion rendered by Scientist Shantanu Das, Senior Scientist B. B. Biswas, Head Reactor Control Division, Bhaba Atomic Research Centre, Mumbai and my former colleague Dr. Subir Das, Department of Mathematics, Institute of Technology, Banaras Hindu University. This is not out of place to acknowledge the effort of my Ph.D. Scholar student and M.Sc. students for their help to write this book.

I, also, express my sincere gratitude to the Director of National Institute of Technology, Rourkela for his kind cooperation in this regard. I received considerable assistance from my colleagues in the Department of Mathematics, National Institute of Technology, Rourkela.

I wish to express my sincere thanks to several people involved in the preparation of this book.

Moreover, I am especially grateful to the Springer Publishing Company for their cooperation in all aspects of the production of this book.

Last, but not the least, special mention should be made of my parents and my beloved wife, Lopamudra for their patience, unequivocal support, and encouragement throughout the period of my work.

I look forward to receive comments and suggestions on the work from students, teachers, and researchers.

Santanu Saha Ray

Contents

About the Author

Dr. S. Saha Ray is currently working as an Associate Professor at the Department of Mathematics, National Institute of Technology, Rourkela, India. Dr. Saha Ray completed his Ph.D. in 2008 from Jadavpur University, India. He received his MCA degree in the year 2001 from Bengal Engineering College, Sibpur, Howrah, India. He completed his M.Sc. in Applied Mathematics at Calcutta University in 1998 and B.Sc. (Honors) in Mathematics at St. Xavier's College, Kolkata, in 1996. Dr. Saha Ray has about 12 years of teaching experience at undergraduate and postgraduate levels. He also has more than 10 years of research experience in various field of Applied Mathematics. He has published several research papers in numerous fields and various international journals of repute like Transaction ASME Journal of Applied Mechanics, Annals of Nuclear Energy, Physica Scripta, Applied Mathematics and Computation, and so on. He is a member of the Society for Industrial and Applied Mathematics (SIAM) and American Mathematical Society (AMS). He was the Principal Investigator of the BRNS research project granted by BARC, Mumbai. Currently, he is acting as Principal Investigator of a research Project financed by DST, Govt. of India. It is not out of place to mention that he had been invited to act as lead guest editor in the journal entitled *International Journal of Differential equations* of Hindawi Publishing Corporation, USA.

Chapter 1
Introduction to Graphs

1.1 Definitions of Graphs

A graph $G = (V(G), E(G))$ or $G = (V, E)$ consists of two finite sets. $V(G)$ or V, the vertex set of the graph, which is a non-empty set of elements called vertices and $E(G)$ or E, the edge set of the graph, which is a possibly empty set of elements called edges, such that each edge e in E is assigned as an unordered pair of vertices (u, v), called the end vertices of e.

Order and size: We define $|V| = n$ to be the *order* of G and $|E| = m$ to be the *size* of G.

Self-loop and parallel edges: The definition of a graph allows the possibility of the edge e having identical end vertices. Such an edge having the same vertex as both of its end vertices is called a *self-loop* (or simply a loop).

Edge e_1 in Fig. 1.1b is a self-loop. Also, note that the definition of graph allows that more than one edge is associated with a given pair of vertices, for example, edges e_4 and e_5 in Fig. 1.1b. Such edges are referred to as *parallel edges*.

Simple graph: A graph, that has neither self-loops nor parallel edges, is called a *simple graph*. An example of a simple graph is given in Fig. 1.1a.

Multigraph: A multigraph G is an ordered pair $G = (V, E)$ with V a set of *vertices* or *nodes* and E a multiset of unordered pairs of vertices called *edges*. An example of a multigraph is given in Fig. 1.1b.

Finite and Infinite graph: A graph with a finite number of vertices as well as finite number of edges is called a finite graph; otherwise it is an infinite graph as shown in Fig. 1.1c.

S. Saha Ray, *Graph Theory with Algorithms and its Applications*,
DOI: 10.1007/978-81-322-0750-4_1, © Springer India 2013

Fig. 1.1 **a** Simple graph,
b multigraph, and **c** infinite
graph

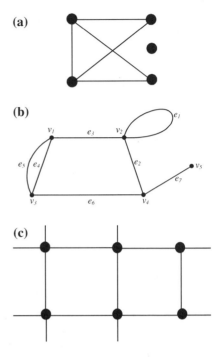

1.2 Some Applications of Graphs

Graph theory has a very wide range of applications in engineering, in physical, and biological sciences, and in numerous other areas.

Königsberg Bridge Problem: The Königsberg Bridge Problem is perhaps the best known example in graph theory. It was a long-standing problem until solved by Euler in 1736 by means of a graph. Euler wrote the first research paper in graph theory and then became the originator of the theory of graphs. The problem is depicted in Fig. 1.2.

The islands C and D formed by the river in Königsberg were connected to each other and to the banks A and B with seven bridges, as shown in Fig. 1.2. The problem was to start at any of the four land areas of the city A, B, C, and D walk over each of the seven bridges exactly once and return to the starting point. Euler represented this situation by means of a graph in Fig. 1.3. The vertices represent the land areas and the edges represent the bridges.

Graph theory was born in 1736 with Euler's famous graph in which he solved the Königsberg Bridge Problem. If some closed walk in a graph contains all the edges of the graph exactly once then (the walk is called an *Euler line* and) the graph is an *Euler graph*.

Remarks A given connected graph G is an *Euler graph* if and only if all the vertices of G are of even degree.

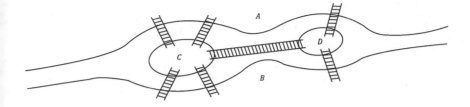

Fig. 1.2 Pictorial representation of Königsberg bridge problem

Fig. 1.3 A graph
representing Königsberg
bridge problem

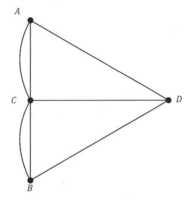

Now looking at the graph of the Königsberg Bridges, we find that not all its vertices are of even degree. Hence, it is not an Euler graph. Thus, it is not possible to walk over each of the seven bridges exactly once and return to the starting point.

Shortest Path Problem: A company has branches in each of six cities where cities are C_1, C_2, C_3, C_4, C_5, and C_6. The airfare for a direct flight from C_i to C_j is given by the (i,j)th entry of the following matrix (where ∞ indicates that there is no direct flight). For example, the fare from C_1 to C_4 is USD 50 and from C_2 to C_3 is USD 15.

$$
\begin{array}{c}
\quad \\
C_1 \\
C_2 \\
C_3 \\
C_4 \\
C_5 \\
C_6
\end{array}
\begin{array}{c}
\begin{array}{cccccc}
C_1 & C_2 & C_3 & C_4 & C_5 & C_6
\end{array} \\
\left[
\begin{array}{cccccc}
0 & 50 & \infty & \infty & \infty & 10 \\
50 & 0 & 15 & 20 & \infty & 25 \\
\infty & 15 & 0 & 10 & \infty & \infty \\
\infty & 20 & 10 & 0 & 10 & 25 \\
\infty & \infty & \infty & 10 & 0 & 55 \\
10 & 25 & \infty & 25 & 55 & 0
\end{array}
\right]
\end{array}
$$

The company is interested in computing a table of cheapest fares between pairs of cities. We can represent the situation by a weighted graph (Fig. 1.4). The problem can be solved using Dijkstra's algorithm.

Fig. 1.4 The weighted graph
representing airfares for
direct flights between six
cities

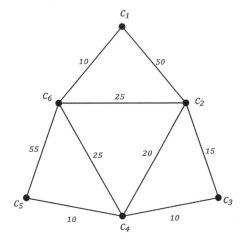

1.3 Incidence and Degree

When a vertex v_i is an end vertex of some edge e_j, v_i, and e_j are said to be incident
with (to or on) each other.

Fig. 1.5 A graph
(multigraph) with five
vertices and seven edges

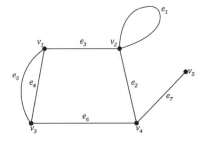

A graph with five vertices and seven edges is shown in Fig. 1.5. Edges e_2, e_6,
and e_7 are incident with vertex v_4.

Adjacent: Two nonparallel edges are said to be adjacent if they are incident on
a common vertex. For example, e_2 and e_7 are adjacent. Similarly, two vertices are
said to be adjacent if they are the end vertices of the same edge. In Fig. 1.5, v_4 and
v_5 are adjacent, but v_1 and v_4 are not.

Degree: Let v be a vertex of the graph G. The degree $d(v)$ of v is the number of
edges of G incident with v, counting each self-loop twice. The *minimum degree*
and the *maximum degree* of a graph G are denoted by $\delta(G)$ and $\Delta(G)$, respectively.
For example, in Fig. 1.5, $d(v_1) = 3 = d(v_3) = d(v_4)$, $d(v_2) = 4$ and $d(v_5) = 1$

$$d(v_1) + d(v_2) + \ldots\ldots + d(v_5) = 14 = \text{twice the number of edges.}$$

Theorem 1.1 *For any graph G with e edges and n vertices v_1, v_2, v_3...... v_n* $\sum_{i=1}^{n} d(v_i) = 2e$.

Proof Each edge, since it has two end vertices, contributes precisely two to the sum of the degrees of all vertices in G. When the degrees of the vertices are summed each edge is counted twice. □

Odd and even vertices: A vertex of a graph is called odd or even depending on whether its degree is odd or even.

In the graph of Fig. 1.5, there is an even number of odd vertices.

Theorem 1.2 (Handshaking lemma) *In any graph G, there is an even number of odd vertices.*

Proof If we consider the vertices with odd and even degrees separately, the equation
$\sum_{i=1}^{n} d(v_i) = 2e$ can be expressed as equation

$$\sum_{i=1}^{n} d(v_i) = \sum_{even} d(v_j) + \sum_{odd} d(v_k)$$

Let W be the set of odd vertices of G, and let U be the set of even vertices of G. Then for each $u \in U$, $d(u)$ is even and so $\sum_{u \in U} d(u)$, being a sum of even numbers, is even.

However,

$$\sum_{u \in U} d(u) + \sum_{w \in W} d(w) = \sum_{v \in V} d(v) = 2e, \text{ by Theorem 1.1}$$

Thus,

$$\sum_{w \in W} d(w) = 2e - \sum_{u \in U} d(u), \text{ is even. (being the difference of two even numbers)}$$

As all the terms in $\sum_{w \in W} d(w)$ are odd and their sum is even, there must be an even number of odd vertices. □

Isolated vertex: A vertex having no incident edge is called an *isolated vertex*. Figure. 1.1a has an isolated vertex.

Pendant vertex: A vertex of degree one is called a *pendant vertex*. In Fig. 1.1b, vertex v_5 is a pendant vertex.

Null graph: If $E = \emptyset$, in a graph $G = (V, E)$, then such a graph without any edges is called a *null graph*.

1.4 Isomorphism

A graph $G_1 = (V_1, E_1)$ is said to be isomorphic to the graph $G_2 = (V_2, E_2)$ if there is a one-to-one correspondence between the vertex sets V_1 and V_2 and a one-to-one correspondence between the edge sets E_1 and E_2 in such a way that if e_1 is an edge with end vertices u_1 and v_1 in G_1 then the corresponding edge e_2 in G_2 has its end vertices u_2 and v_2 in G_2 which corresponds to u_1 and v_1, respectively. Such a pair of correspondence is called a *graph isomorphism*.

In other words, two graphs G and G' are said to be isomorphic if there is a one-to-one correspondence between their vertices and between their edges such that the incidence relationship is preserved.

Example 1.1 Show that the following two graphs in Fig. 1.6 are isomorphic.

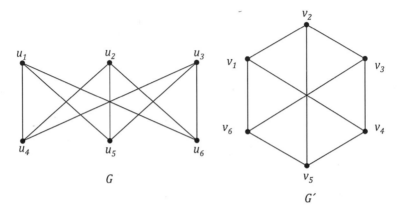

Fig. 1.6 Two isomorphic graphs G and G'

Solution:
We see that both the graphs G and G' have equal number of vertices and edges. The vertex corresponds are given below:

$u_1 \leftrightarrow v_1,\ u_2 \leftrightarrow v_3,\ u_3 \leftrightarrow v_5,\ u_4 \leftrightarrow v_2,\ u_5 \leftrightarrow v_4,\ u_6 \leftrightarrow v_6$ or $u_5 \leftrightarrow v_6,\ u_6 \leftrightarrow v_4$.

Hence, the two graphs are isomorphic.

Example 1.2 Check whether the graphs in Fig. 1.7 are isomorphic.

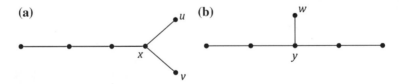

Fig. 1.7 Two non-isomorphic graphs

Solution:
The graphs in Fig. 1.7a and b are not isomorphic. If the graph 1.7a were to be isomorphic to the one in 1.7b, vertex x must correspond to y, because there are no other vertices of degree three. Now in 1.7b, there is only one pendant vertex w adjacent to y, while in 1.7a there are two pendant vertices u and v adjacent to x.

1.5 Complete Graph

A complete graph is a simple graph in which each pair of distinct vertices is joined by an edge. In other words, a simple graph in which there exists an edge between every pair of vertices is called a complete graph. If the complete graph has vertices $v_1, v_2,....v_n$, then the edge set can be given by

$$E = \{(v_i, v_j) : v_i \neq v_j; \quad i,j = 1,2,3...n\}$$

It follows that the graph has $n(n-1)/2$ edges (since there are $n-1$ edges incident with each of the n vertices, so a total of $n(n-1)$, but divide by 2 since $(v_j, v_i) = (v_i, v_j)$).

Corollary *The maximum number of edges in a simple graph with n vertices is* $n(n-1)/2$. *Given any two complete graphs with the same number of vertices, n, then they are isomorphic.*

The complete graph of n vertices is denoted by K_n.

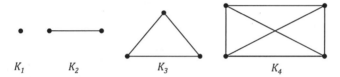

K_1 K_2 K_3 K_4

Fig. 1.8 Complete graphs K_1, K_2, K_3, and K_4

Figure 1.8 shows K_1, K_2, K_3 and K_4.
Trivial graph: An empty (or *trivial*) graph is a graph with no edges.

1.6 Bipartite Graph

Definition Let G be a graph. If the vertex set V of G can be partitioned into two non-empty subsets X and Y (*i.e.,* $X \cup Y = V$ *and* $X \cap Y = \emptyset$) in such a way that, each edge of G has one end in X and other end in Y, then G is called *bipartite*. The partition $V = X \cup Y$ is called a *bipartition* of G.

Figures 1.9 and 1.10 cite examples of Bipartite graphs.

1.6.1 Complete Bipartite Graph

Definition A complete Bipartite graph is a simple bipartite graph G, with bipartition $V = X \cup Y$ in which every vertex in X is adjacent to every vertex of Y. If X has m vertices and Y has n vertices, such a graph is denoted by $K_{m,n}$.

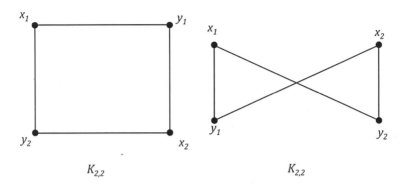

Fig. 1.9 Complete bipartite graph $K_{2,2}$

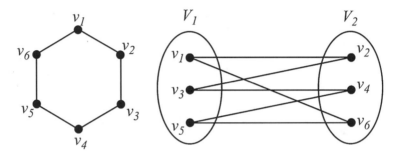

Fig. 1.10 A bipartite graph

Corollary *Any complete bipartite graph with a bipartition into two sets of m and n vertices is isomorphic to $K_{m,n}$.*

Since each of the *m* vertices in the partition set *X* of $K_{m,n}$ is adjacent to each of the *n* vertices in the partition set *Y*, $K_{m,n}$ has *m* * *n* edges.

Figure 1.11 shows complete bipartite graphs.

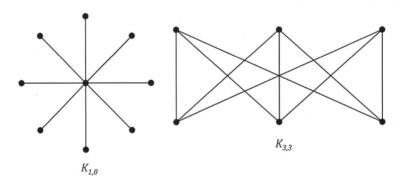

Fig. 1.11 Complete bipartite graphs $K_{1,8}$ and $K_{3,3}$

*k***-Regular**: If for some positive integer $k, d(v) = k$ for every vertex v of the graph G, then G is called *k-regular*.

A regular graph is one that is *k-regular* for some k.

For example, the graph $K_{2,2}$ shown in Fig. 1.9 is *2-regular*. The complete graph K_n is $(n - 1)$-regular. The complete bipartite graph $K_{n,n}$ on **2n** vertices is *n-regular*.

1.7 Directed Graph or Digraph

A digraph (or a directed graph) $G = (V_G, E_G)$ consists of the two sets:

1. A vertex set V_G, nonempty set, whose elements are called vertices or nodes.
2. An edge set or arc set E_G, possibly empty set, whose elements are called directed edges or arcs, such that each directed edge in E_G is assigned an order pair of vertices (u, v), i.e., $E_G \subseteq V_G \times V_G$.

For $u, v \in V_G$, an arc or a directed edge $e = (u, v) \in V_G$ is denoted by uv and implies that e is directed from u to v. Here, u is the initial vertex and v is the terminal vertex. Also, we say that e joins u to v; e is incident with u and v; e is incident from u and e is incident to v; and u is adjacent to v and v is adjacent from u. For example, Fig. 1.12 shows a directed graph or digraph.

In-degree and Out-degree: The in-degree and the out-degree of a vertex are defined as follows:

1. In a digraph G, the number of edges incident out of a vertex v is called the out-degree of v. It is denoted by $\text{degree}^+(v)$ or $d^+(v)$.
2. In a digraph G, the number of edges incident into a vertex v is called the in-degree of v. It is denoted by $\text{degree}^-(v)$ or $d^-(v)$.

The total degree (or simply degree) of v is $d(v) = \text{degree}^+(v) + \text{degree}^-(v)$.
In this case, we have the following Handshaking Lemma.

Lemma 1.1 *Let G be a digraph. Then*

$$\sum_{v \in G} \text{degree}^+(v) = |E_G| = \sum_{v \in G} \text{degree}^-(v)$$

Example 1.3 Find the in-degree and out-degree of each vertex of the following directed graph. Also, verify that the sum of the in-degrees (or the out-degrees) equals the number of edges.

Fig. 1.12 A directed graph or digraph

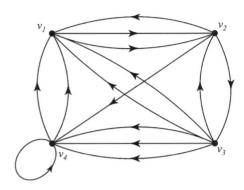

Solution:
For the graph G in Fig. 1.12

$$\text{degree}^+(v_1) = 2 \quad \text{degree}^-(v_1) = 5$$

$$\text{degree}^+(v_2) = 3 \quad \text{degree}^-(v_2) = 3$$

$$\text{degree}^+(v_3) = 6 \quad \text{degree}^-(v_3) = 1$$

$$\text{degree}^+(v_4) = 3 \quad \text{degree}^-(v_4) = 5$$

Here, we see that

$$\sum_{v \in G} \text{degree}^+(v) = \sum_{v \in G} \text{degree}^-(v) = 14 = \text{the number of edges of } G.$$

Chapter 2
Subgraphs, Paths, and Connected Graphs

2.1 Subgraphs and Spanning Subgraphs (Supergraphs)

Subgraph: Let H be a graph with vertex set $V(H)$ and edge set $E(H)$, and similarly let G be a graph with vertex set $V(G)$ and edge set $E(G)$. Then, we say that H is a subgraph of G if $V(H) \subseteq V(G)$ and $E(H) \subseteq E(G)$. In such a case, we also say that G is a supergraph of H.

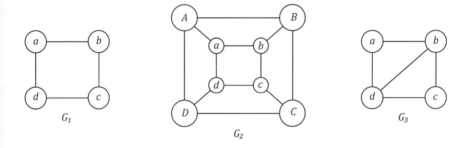

$$G_1 \subseteq G_2, \ G_1 \subseteq G_3 \text{ but } G_3 \nsubseteq G_2.$$

Fig. 2.1 G_1 is a subgraph of G_2 and G_3

In Fig. 2.1, G_1 is a subgraph of both G_2 and G_3 but G_3 is not a subgraph of G_2. Any graph isomorphic to a subgraph of G is also referred to as a subgraph of G. If H is a subgraph of G then we write $H \subseteq G$. When $H \subseteq G$ but $H \neq G$, i.e., $V(H) \neq V(G)$ or $E(H) \neq E(G)$, then H is called a proper subgraph of G.

Spanning subgraph (or Spanning supergraph): A *spanning subgraph* (or *spanning supergraph*) of G is a *subgraph* (or *supergraph*) H with $V(H) = V(G)$, i.e. H and G have exactly the same vertex set.

It follows easily from the definitions that any simple graph on n vertices is a subgraph of the complete graph K_n. In Fig. 2.1, G_1 is a proper spanning subgraph of G_3.

S. Saha Ray, *Graph Theory with Algorithms and its Applications*,
DOI: 10.1007/978-81-322-0750-4_2, © Springer India 2013

2.2 Operations on Graphs

The *union* of two graphs $G_1 = (V_1, E_1)$ and $G_2 = (V_2, E_2)$ is another graph $G_3 = (V_3, E_3)$ denoted by $G_3 = G_1 \cup G_2$, where vertex set $V_3 = V_1 \cup V_2$ and the edge set $E_3 = E_1 \cup E_2$.

The *intersection* of two graphs G_1 and G_2 denoted by $G_1 \cap G_2$ is a graph G_4 consisting only of those vertices and edges that are in both G_1 and G_2.

The *ring sum* of two graphs G_1 and G_2, denoted by $G_1 \oplus G_2$, is a graph consisting of the vertex set $V_1 \cup V_2$ and of edges that are either in G_1 or G_2, but not in both.

Figure 2.2 shows union, intersection, and ring sum on two graphs G_1 and G_2.

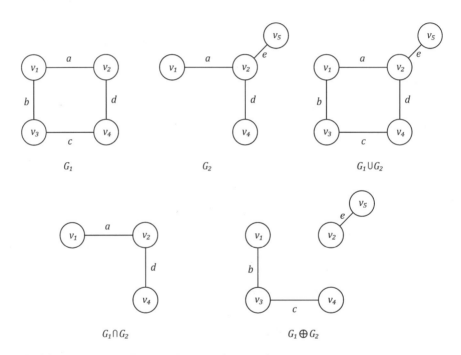

Fig. 2.2 Union, intersection, and ring sum of two graphs

Three operations are *commutative*, i.e.,

$$G_1 \cup G_2 = G_2 \cup G_1, \quad G_1 \cap G_2 = G_2 \cap G_1, \quad G_1 \oplus G_2 = G_2 \oplus G_1$$

If G_1 and G_2 are edge disjoint, then $G_1 \cap G_2$ is a null graph, and $G_1 \oplus G_2 = G_1 \cup G_2$. If G_1 and G_2 are vertex disjoint, then $G_1 \cap G_2$ is empty.

For any graph G, $G \cap G = G \cup G = G$ and $G \oplus G = $ *a null graph*.

If g is a subgraph of G, i.e., $g \subseteq G$, then $G \oplus g = G - g$, and is called a *complement* of g in G.

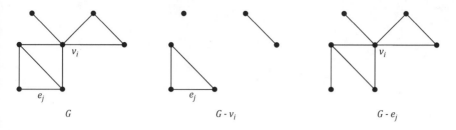

Fig. 2.3   Vertex deletion and edge deletion from a graph G

Decomposition: A graph G is said to be decomposed into two subgraphs G_1 and G_2, if $G_1 \cup G_2 = G$ and $G_1 \cap G_2$ is a *null graph*.

Deletion: If v_i is a vertex in graph G, then $G - v_i$ denotes a subgraph of G obtained by deleting v_i from G. Deletion of a vertex always implies the deletion of all edges incident on that vertex. If e_j is an edge in G, then $G - e_j$ is a subgraph of G obtained by deleting e_j from G. Deletion of an edge does not imply deletion of its end vertices. Therefore, $G - e_j = G \oplus e_j$ (Fig. 2.3).

Fusion: A pair of vertices a, b in a graph G are said to be *fused* if the two vertices are replaced by a single new vertex such that every edge, that was incident on either a or b or on both, is incident on the new vertex. Thus, fusion of two vertices does not alter the number of edges, but reduces the number of vertices by one (Fig. 2.4).

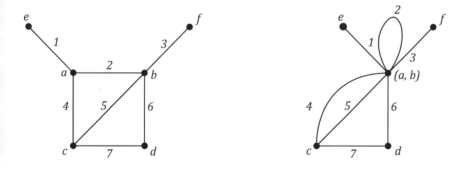

Fig. 2.4   Fusion of two vertices a and b

Induced subgraph: A subgraph $H \subseteq G$ is an induced subgraph, if $E_H = E_G \cap E(V_H)$. In this case, H is induced by its set V_H of vertices. In an induced subgraph $H \subseteq G$, the set E_H of edges consists of all $e \in E_G$, such that $e \in E(V_H)$. To each nonempty subset $A \subseteq V_G$, there corresponds a unique induced subgraph $G[A] = (A, E_G \cap E(A))$ (Fig. 2.5).

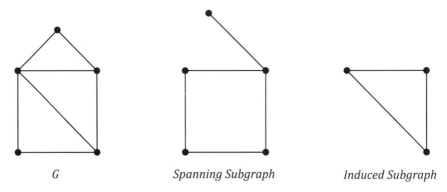

| G | *Spanning Subgraph* | *Induced Subgraph* |

Fig. 2.5 Spanning subgraph and induced subgraph of a graph G

Trivial graph: A graph $G = (V, E)$ is trivial, if it has only one vertex. Otherwise G is nontrivial.

Discrete graph: A graph is called discrete graph if $E_G = \phi$.

Stable: A subset $X \subseteq V_G$ is stable, if $G[X]$ is a discrete graph.

2.3 Walks, Trails, and Paths

Walk: A *walk* in a graph G is a finite sequence

$$W \equiv v_0 e_1 v_1 e_2 \cdots v_{k-1} e_k v_k$$

whose terms are alternately vertices and edges such that for $1 \leq i \leq k$, the edge e_i has ends v_{i-1} and v_i.

Thus, each edge e_i is immediately preceded and succeeded by the two vertices with which it is incident. We say that W is a $v_0 - v_k$ walk or a walk from v_0 to v_k.

Origin and terminus: The vertex v_0 is the *origin* of the walk W, while v_k is called the *terminus* of W. v_0 and v_k need not be distinct.

The vertices $v_1, v_2, \ldots, v_{k-1}$ in the above walk W are called its *internal vertices*. The integer k, the number of edges in the walk, is called the *length of W*, denoted by $|W|$.

In a walk W, there may be repetition of vertices and edges.

Trivial walk: A *trivial walk* is one containing no edge. Thus for any vertex v of G, $W \equiv v$ gives a trivial walk. It has length 0.

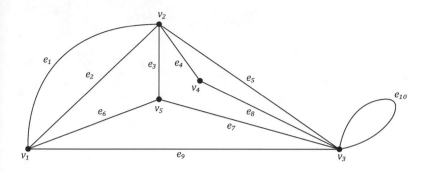

Fig. 2.6 A graph with five vertices and ten edges

In Fig. 2.6, $W_1 \equiv v_1 e_1 v_2 e_5 v_3 e_{10} v_3 e_5 v_2 e_3 v_5$ and $W_2 \equiv v_1 e_1 v_2 e_1 v_1 e_1 v_2$ are both walks of length 5 and 3, respectively, from v_1 to v_5 and from v_1 to v_2, respectively.

Given two vertices u and v of a graph G, a u–v walk is called *closed* or *open*, depending on whether $u = v$ or $u \neq v$.

Two walks W_1 and W_2 above are both open, while $W_3 \equiv v_1 v_5 v_2 v_4 v_3 v_1$ is closed in Fig. 2.6.

Trail: If the edges $e_1, e_2, \ldots, e_k$ of the walk $W \equiv v_0 e_1 v_1 e_2 v_2 \cdots\cdots e_k v_k$ are distinct then W is called a *trail*. In other words, a trail is a walk in which no edge is repeated. W_1 and W_2 are not trails, since for example e_5 is repeated in W_1, while e_1 is repeated in W_2. However, W_3 is a trail.

Path: If the vertices $v_0, v_1, \ldots, v_k$ of the walk $W \equiv v_0 e_1 v_1 e_2 v_2 \cdots e_k v_k$ are distinct then W is called a *path*. Clearly, any two paths with the same number of vertices are isomorphic.

A path with n vertices will sometimes be denoted by P_n.

Note that P_n has length $n - 1$.

In other words, *a path is a walk in which no vertex is repeated*. Thus, in a path no edge can be repeated either, so a every path is a trail. Not every trail is a path, though. For example, W_3 is not a path since v_1 is repeated. However, $W_4 \equiv v_2 v_4 v_3 v_5 v_1$ is a path in the graph G as shown in Fig. 2.6.

2.4 Connected Graphs, Disconnected Graphs, and Components

Connected vertices: A vertex u is said to be *connected* to a vertex v in a graph G if there is a path in G from u to v.

Connected graph: A graph G is called *connected* if every two of its vertices are connected.

Disconnected graph: A graph that is not connected is called *disconnected*.

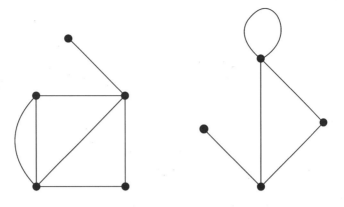

Fig. 2.7 A disconnected graph with two components

It is easy to see that a disconnected graph consists of two or more *connected graphs*. Each of these connected subgraphs is called a component. Figure 2.7 shows a disconnected graph with two components.

Theorem 2.1 *A graph G is disconnected iff its vertex set V can be partitioned into two non-empty, disjoint subsets V_1 and V_2 such that there exists no edge in G whose one end vertex is in subset V_1 and the other in subset V_2.*

Proof Suppose that such a partitioning exists. Consider two arbitrary vertices *a* and *b* of *G*, such that $a \in V_1$ and $b \in V_2$. No path can exist between vertices *a* and *b*; otherwise there would be at least one edge whose one end vertex would be in V_1 and the other in V_2. Hence, if a partition exists, *G* is not connected.

Conversely, let *G* be a disconnected graph. Consider a vertex *a* in *G*. Let V_1 be the set of all vertices that are connected by paths to *a*. Since *G* is disconnected, V_1 does not include all vertices of *G*. The remaining vertices will form a (non-empty) set V_2. No vertex in V_1 is connected to any vertex in V_2 by path. Hence the partition exists.   □

Theorem 2.2 *If a graph (connected or disconnected) has exactly two vertices of odd degree, there must be a path joined by these two vertices.*

Proof Let *G* be a graph with all even vertices except vertices v_1 and v_2, which are odd. From Handshaking lemma, which holds for every graph and therefore for every component of a disconnected graph, no graph can have an odd number of odd vertices. Therefore, in graph *G*, v_1 and v_2 must belong to the same component, and hence there must be a path between them.   □

Theorem 2.3 *A simple graph with n vertices and k components can have at most $(n - k)(n - k + 1)/2$ edges.*

Proof Let the number of vertices in each of the k components of a graph G be $n_1, n_2, \ldots, n_k$. Thus, we have

$$n_1 + n_2 + \cdots + n_k = n$$

where $n_i \geq 1 \quad for \quad i = 1, 2, \ldots, k.$

Now, $\sum_{i=1}^{k} (n_i - 1) = n - k$

$$\Rightarrow \left(\sum_{i=1}^{k} (n_i - 1) \right)^2 = n^2 + k^2 - 2nk$$

$$\Rightarrow [(n_1 - 1) + (n_2 - 1) + \cdots + (n_k - 1)]^2 = n^2 + k^2 - 2nk$$

$$\Rightarrow \sum_{i=1}^{k} (n_i - 1)^2 + 2 \sum_{i,j=1, i \neq j}^{k} (n_i - 1)(n_j - 1) = n^2 + k^2 - 2nk$$

$$\Rightarrow \sum_{i=1}^{k} (n_i)^2 - 2 \sum_{i=1}^{k} n_i + k + 2 \sum_{i,j=1, i \neq j}^{k} (n_i - 1)(n_j - 1) = n^2 + k^2 - 2nk$$

$$\Rightarrow \sum_{i=1}^{k} n_i^2 - 2n + k + 2 \sum_{i,j=1, i \neq j}^{k} (n_i - 1)(n_j - 1) = n^2 + k^2 - 2nk$$

$$\Rightarrow \sum_{i=1}^{n} n_i^2 + 2 \sum_{i,j=1, i \neq j}^{k} (n_i - 1)(n_j - 1) = n^2 + k^2 - 2nk + 2n - k.$$

Since each $(n_i - 1) \geq 0.$

$$\sum_{i=1}^{n} n_i^2 \leq n^2 + k^2 - 2nk + 2n - k = n^2 + k(k - 2n) - (k - 2n)$$

$$= n^2 - (k - 1)(2n - k)$$

Now, the maximum number of edges in the ith component of G is $n_i(n_i - 1)/2$. Since the maximum number of edges in a simple graph with n vertices is $n(n-1)/2$ therefore, the maximum number of edges in G is

$$\frac{1}{2} \sum_{i=1}^{k} n_i(n_i - 1) = \frac{1}{2} \sum_{i=1}^{n} n_i^2 - \frac{n}{2}$$

$$\leq \frac{1}{2} [n^2 - (k - 1)(2n - k)] - \frac{n}{2}$$

$$= \frac{1}{2} [n^2 - 2nk + 2n + k^2 - k - n]$$

$$= \frac{1}{2} \left[(n - k)^2 + (n - k) \right]$$

$$= \frac{1}{2} (n - k)(n - k + 1) \qquad \square$$

2.5 Cycles

Cycle: A nontrivial closed trail in a graph G is called a cycle if its origin and internal vertices are distinct. In detail, the closed trail

$C \equiv v_1 v_2 \cdots v_n v_1$ is a cycle if

1. C has at least one edge and
2. $v_1, v_2, \ldots, v_n$ are n distinct vertices.

k-Cycle: A cycle of length k, , i.e., with k edges, is called a k-cycle. A k-cycle is called odd or even depending on whether k is odd or even.

Figure 2.8 cites C_3, C_4, C_5, and C_6. A 3-cycle is often called a triangle. Clearly, any two cycles of the same length are isomorphic.

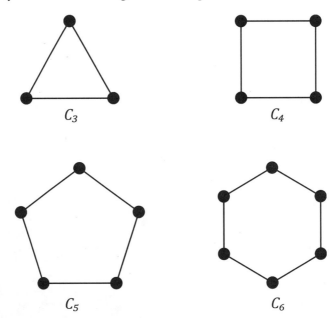

Fig. 2.8 Cycles C_3, C_4, C_5 and C_6

An *n*-cycle, i.e., a cycle with n vertices, will sometimes be denoted by C_n.

In Fig. 2.9, $C \equiv v_1 v_2 v_3 v_4 v_1$, is a 4-cycle and $T \equiv v_1 v_2 v_5 v_3 v_4 v_5 v_1$ is a non-trivial closed trail which is not a cycle (because v_5 occurs twice as an internal vertex) and $C' \equiv v_1 v_2 v_5 v_1$ is a triangle.

Fig. 2.9 A graph containing 3-cycles and 4-cycles

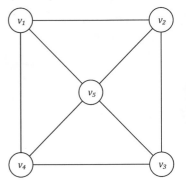

Fig. 2.10 A 2-cycle

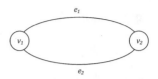

Note that, a loop is just a 1-cycle. Also, given parallel edges e_1 and e_2 in Fig. 2.10 with distinct end vertices v_1 and v_2, we can find the cycle $v_1e_1v_2e_2v_1$ of length 2. Conversely, the two edges of any cycle of length 2 are a pair of parallel edges.

Theorem 2.4 *Given any two vertices u and v of a graph G, every u–v walk contains a u–v path.*

Proof We prove the statement by induction on the length l of a u–v walk W.

Basic step: $l = 0$, having no edge, W consists of a single vertex ($u = v$). This vertex is a u–v path of length 0.

Induction step: $l \geq 1$. We suppose that the claim holds for walks of length less than l. If W has no repeated vertex, then its vertices and edges form a u–v path. If W has a repeated vertex w, then deleting the edges and vertices between appearances of w (leaving one copy of w) yields a shorter u-v walk W' contained in W. By the induction hypothesis, W' contains a u–v path P, and this path P is contained in W (Fig. 2.11). This proves the theorem.   □

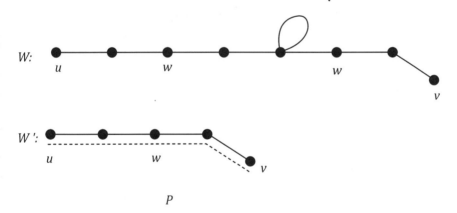

Fig. 2.11 A walk W and a shorter walk W' of W containing a path P

Theorem 2.5 *The minimum number of edges in a connected graph with n vertices is $n − 1$.*

Proof Let m be the number of edges of such a graph. We have to show $m \geq n − 1$. We prove this by method of induction on m. If $m = 0$ then obviously $n = 1$ (otherwise G will be disconnected). Clearly, then $m \geq n − 1$. Let the result be true for $m = 0, 1, 2, 3, \ldots, k$. We shall show that the result is true for $m = k + 1$. Let G be a graph with $k + 1$ edges. Let e be an edge of G. Then the subgraph $G − e$ has

k edges and n number of vertices. If $G - e$ is also connected then by our hypothesis $k \geq n - 1$, i.e., $k + 1 \geq n > n - 1$.

If $G - e$ is disconnected then it would have two connected components. Let the two components have k_1, k_2 number of edges and n_1, n_2 number of vertices, respectively. So, by our hypothesis, $k_1 \geq n_1 - 1$ and $k_2 \geq n_2 - 1$. These two imply that $k_1 + k_2 \geq n_1 + n_2 - 2$, i.e., $k \geq n - 2$ (since, $k_1 + k_2 = k$, $n_1 + n_2 = n$), i.e., $k + 1 \geq n - 1$.

Thus, the result is true for $m = k + 1$. $\qquad\qquad\qquad\qquad\qquad\qquad\qquad\qquad\qquad\qquad$ □

Theorem 2.6 *A graph G is bipartite if and only if it has no odd cycles.*

Proof Necessary condition:

Let G be a bipartite graph with bipartition (X, Y), i.e., $V = X \cup Y$.

For any cycle $C : v_1 \rightarrow v_2 \cdots \rightarrow v_{k+1}(= v_1)$ of length $k, v_1 \in X \Rightarrow v_2 \in Y, v_3 \in X \Rightarrow v_4 \in Y \cdots \Rightarrow v_{2m} \in Y \Rightarrow v_{2m+1} \in X$. Consequently, $k + 1 = 2m + 1$ is odd and $k = |C|$ is even. Hence, G has no odd cycle.

Sufficient condition:

Suppose that, all the cycles in G are even, i.e., G be a graph with no odd cycle.

To show: G is a bipartite graph. It is sufficient to prove this theorem for the connected graph only.

Let us assume that G is connected. Let $v \in G$ be an arbitrary chosen vertex. Now, we define,

$$X = \{x | d_G(v, x) \text{ is even}\},$$

i.e., X is the set of all vertices x of G with the property that any shortest $v - x$ path of G has even length and $Y = \{y | d_G(v, y) \text{ is odd}\}$, i.e., Y is the set of all vertices y of G with the property that any shortest $v - y$ path of G has odd length.

Here,

$$d_G(u, v) = \text{shortest distance from the vertex } u \text{ to the vertex } v$$

$$= \min\left\{k : u \xrightarrow{k} v\right\}$$

[If the graph G is connected then this shortest distance should be finite, i.e., $d_G(u, v) < \infty$ for $\forall u, v \in G$. Otherwise, G is disconnected]

Then clearly, since the graph G is connected $V = X \cup Y$ and also by definition of distance $X \cap Y = \emptyset$.

Now, we show that $V = X \cup Y$ is a bipartition of G by showing that any edge of G must have one end vertex in X and another in Y.

Suppose that $u, w \in V(G)$ are both either in X or in Y and they are adjacent.

Let $P : v \xrightarrow{*} u$ and $Q : v \xrightarrow{*} w$ be the two shortest paths from v to u and v to w, respectively.

Let x be the last common vertex of the two shortest paths P and Q such that $P = P_1 P_2$ and $Q = Q_1 Q_2$ where $P_2 : x \xrightarrow{*} u$ and $Q_2 : x \xrightarrow{*} w$ are independent (Fig. 2.12).

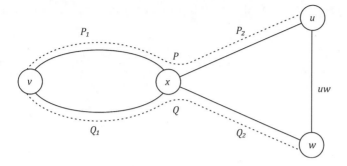

Fig. 2.12 Two shortest paths P and Q

Since P and Q are shortest paths, therefore, $P_1 : v \xrightarrow{*} x$ and $Q_1 : v \xrightarrow{*} x$ are shortest paths from v to x.

Consequently, $|P_1| = |Q_1|$

Now consider the following two cases.

Case 1: $u, w \in X$, then $|P|$ is even and $|Q|$ is even (Also, $|P_1| = |Q_1|$)

Case 2: $u, w \in Y$, then $|P|$ is odd and $|Q|$ is odd (Also, $|P_1| = |Q_1|$)

Therefore, in either case, $|P_2| + |Q_2|$ must be even and so $uw \notin E(G)$. Otherwise, $x \xrightarrow{*} u \to w \xrightarrow{*} x$ would be an odd cycle, which is a contradiction.

Therefore, X and Y are stable subsets of V. This implies (X, Y) is a bipartition of G. Therefore, $G[X]$ and $G[Y]$ are discrete induced subgraphs of G.

Hence, G is a bipartite graph.

If G is disconnected then each cycle of G will belong to any one of the connected components of G say $G_1, G_2, \ldots, G_p$.

If G_i is bipartite with bipartition (X_i, Y_i), then $\big(X_1 \cup X_2 \cup X_3 \cup \cdots \cdots \cup X_p,$ $Y_1 \cup Y_2 \cup \cdots \cdots \cup Y_p\big)$ is a bipartition of G.

Hence, the disconnected graph G is bipartite. $\qquad \square$

Exercises:

1. Show that the following two graphs are isomorphic (Fig. 2.13).

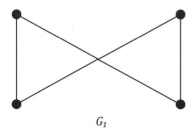

G_1

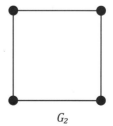

G_2

Fig. 2.13

2. Check whether the following two graphs are isomorphic or not (Fig. 2.14).

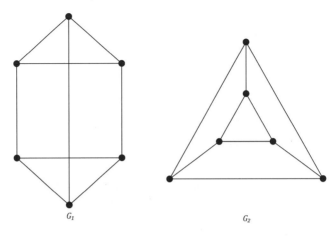

G_1 G_2

Fig. 2.14

3. Show that the following graphs are isomorphic and each graph has the same bipartition (Fig. 2.15).

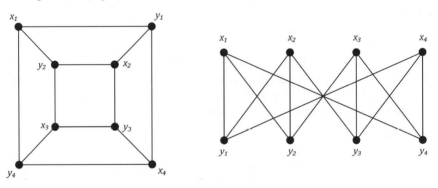

Fig. 2.15

4. What is the difference between a closed trail and a cycle?
5. Are the following graphs isomorphic? (Fig. 2.16).

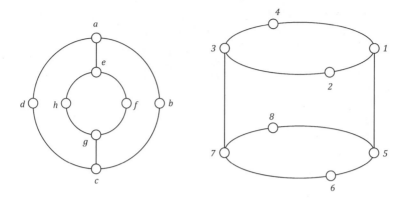

Fig. 2.16

6. Prove that a simple graph having n number of vertices must be connected if it has more than $(n-1)(n-2)/2$ edges.
7. Check whether the following two given graphs G_1 and G_2 are isomorphic or not (Fig. 2.17).

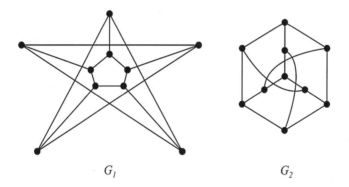

$$G_1 \qquad\qquad G_2$$

Fig. 2.17

8. Prove that the number of edges in a bipartite graph with n vertices is at most $(n^2/2)$.
9. Prove that there exists no simple graph with five vertices having degree sequence $4, 4, 4, 2, 2$.
10. Find, if possible, a simple graph with five vertices having degree sequence $2, 3, 3, 3, 3$.
11. If a simple regular graph has n vertices and 24 edges, find all possible values of n.

12. If $\delta(G)$ and $\Delta(G)$ be the minimum and maximum degrees of the vertices of a graph G with n vertices and e edges, show that

$$\delta(G) \le \frac{2e}{n} \le \Delta(G)$$

13. Show that the minimum number of edges in a simple graph with n vertices is $n - k$, where k is the number of connected components of the graph.

14. Find the maximum number of edges in
    (a) a simple graph with n vertices
    (b) a bipartite graph with bipartition (X, Y) where $|X| = m$ and $|Y| = n$, respectively.

Chapter 3
Euler Graphs and Hamiltonian Graphs

3.1 Euler Tour and Euler Graph

Euler trail: A trail in G is said to be an Euler Trail if it includes all the edges of graph G. Thus a trail is Euler if each edge of G is in the trail exactly once.

Tour: A tour of G is a closed walk of G which includes every edge of G at least once.

Euler tour: An Euler Tour of a graph G is a tour which includes every edge of G exactly once. In other words, a closed Euler Trail is an Euler Tour.

Euler graph: A graph G is called Eulerian or Euler graph if it has an Euler Tour.

For example, the graphs G_1 and G_2 of Fig. 3.1 have an Euler trail and an Euler tour, respectively. In G_1, an Euler trail is given by the sequence of edges $e_1, e_2, e_3, e_4, e_5, e_6, e_7, e_8, e_9, e_{10}$, while in G_2 an Euler tour is given by $e_1, e_2, e_3, e_4, e_5, e_6, e_7, e_8, e_9, e_{10}, e_{11}, e_{12}$.

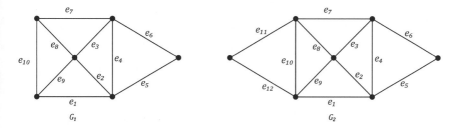

Fig. 3.1 G_1 is not an Euler graph, but G_2 is an Euler graph

In G_1: it has an Euler trail but not Euler tour because it is not closed.

In G_2: all the vertices are even degree. Hence, it is Eulerian which implies it contains the Euler tour. Since G_2 contains Euler Tour so it is Eulerian.

S. Saha Ray, *Graph Theory with Algorithms and its Applications*,
DOI: 10.1007/978-81-322-0750-4_3, © Springer India 2013

Theorem 3.1(Euler Theorem) *A connected graph G is Eulerian (Euler graph) iff every vertex has an even degree.*

Proof Necessary condition:

Let the graph G be Eulerian.

Let $W : u \xrightarrow{*} u$ be an Euler tour and v be any *internal vertex* such that $v \neq u$. Suppose, v appears k times in this Euler tour W. Since every time an edge arrives at v, another edge departs from v, and therefore, $d_G(v) = 2k$ (Even). Also, $d_G(u)$ is 2, since W starts and ends at u.

Hence, the graph G has vertices of all even degree.

Sufficient condition:

Let us assume G is a non-trivial connected graph such that for all vertex $v \in V(G)$, $d_G(v)$ is even.

To show: G is Eulerian.

Let $W = e_1 \ldots e_n : v_0 \xrightarrow{*} v_n$, where $e_i = v_{i-1}v_i$ and W be the largest trail in G. It follows that all $e = v_n w \in E(G)$ are among the edges of W, otherwise We would be the longer than W in G, which is a contradiction. In particular, $v_0 = v_n$,i.e., the trail W is a closed trail. Indeed if $v_0 \neq v_n$ then v_n may appear k times in the trail W, then $d(v_n) = 2(k - 1) + 1 = 2k - 1$ (Odd), which is a contradiction. So W should be closed trail.

If W is not an Euler tour, then since G is connected, there exists an edge $f = v_i u \in E(G)$ for some i, such that f is not in W. Then, $e_{i+1} \ldots e_n e_1 \ldots e_i f$ is a trail in G and it is longer than W. This contradiction to the choice of W proves the claim. So, W is a closed Euler tour. Hence G is a Euler graph. □

Theorem 3.2 *A connected graph has an Euler trail iff it has at most two vertices of odd degree.*

Proof Necessary condition:

Let the graph G has an Euler trail $u \xrightarrow{*} v$. Let w be any vertex which is different from u and v, i.e., $w \neq u, v$. If w is a vertex different from the origin and terminus of the trail, the degree of w is even. Since if w occurs k times then $d(w) = 2k$(even). Thus the only possible odd vertices are the origin and terminus of the trail. If $u(or\ v)$ occurs k times in W, then $d(u) = d(v) = 2(k - 1) + 1$ which is odd.

Hence G has at most two vertices of odd degree.

Sufficient condition:

Let us assume G to be a connected graph and G has at most two vertices of odd degree.

To show: G has an Euler trail.

If G has no odd degree vertices then G has an Euler trail. (just follows from previous Euler theorem). Otherwise, by the *Handshaking Theorem*, every graph has an even number of odd vertices. So, the graph G has exactly two such vertices of odd degree say u and v. Let H be a graph obtained from G by adding a vertex w and the edges uw and vw. So in graph H, every vertex has an even degree. Then, according to Euler theorem H has a Euler tour say $u \xrightarrow{*} v \rightarrow w \rightarrow u$. Here, the beginning part $u \xrightarrow{*} v$ is an Euler trail of G.

Hence the theorem is proved. □

3.2 Hamiltonian Path

Hamiltonian Path: A Hamiltonian path in a graph G is a path which includes every vertex of G.

Hamiltonian Cycle/Circuit: A Hamiltonian cycle in a graph G is a cycle which includes every vertex in G.

Hamiltonian Graph: A graph G is called *Hamiltonian* if it has a Hamiltonian cycle.

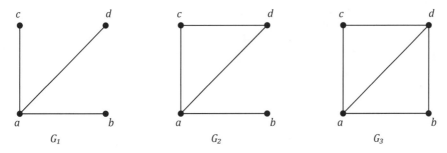

Fig. 3.2 G_1 has no Hamiltonian path, G_2 has a Hamiltonian path, and G_3 has a Hamiltonian cycle

In Fig. 3.2, G_1 has no Hamiltonian path, G_2 has a Hamiltonian path but no Hamiltonian cycle, while G_3 has a Hamiltonian cycle.

It is obvious that each complete graph K_n has a Hamiltonian cycle whenever $n \geq 3$. Consequently, K_n is Hamiltonian for $n \geq 3$. Also, $K_{m,n}$ is Hamiltonian iff $m = n \geq 2$.

3.2.1 Maximal Non-Hamiltonian Graph

A simple graph G is called maximal non-Hamiltonian if it is not Hamiltonian but in addition to it, any edge connecting two nonadjacent vertices forms a Hamiltonian graph.

Theorem 3.3 (Dirac's Theorem 1952) If G is a simple graph with n vertices where $n \geq 3$ and $d(v) \geq {}^n/_2$ for every vertex v of G, then G is Hamiltonian.

Proof We suppose that the result is not true. So, the graph G is Non-Hamiltonian. Then for some value of $n \geq 3$, there is a non-Hamitonian graph in which every vertex has degree at least $n/2$. Any proper spanning supergraph also has every vertex with degree at least $n/2$ because any proper spanning supergraph can be obtained by introducing more edges in G. Thus, there will be a Maximal Non-Hamiltonian graph of G with n vertices and $d(v) \geq {}^n/_2$ for every vertex v in G. But the graph G cannot be complete, since if G is complete graph K_n then it would be a Hamiltonian graph (for $n \geq 3$). Therefore, there are two nonadjacent vertices u and

v in G. Let $G + uv$ be the supergraph of G obtained by introducing an edge uv. Then, since G is Maximal Non-Hamiltonian graph, $G + uv$ must be a Hamiltonian graph. Also, if C is a Hamiltonian cycle of $G + uv$ then C must contain the edge uv. Otherwise it will be a Hamiltonian cycle in G. Thus, choosing such a cycle $C \equiv v_1 v_2 \ldots v_n v_1$, where $v_1 = u$ and $v_n = v$ (the edge $v_n v_1$ is just vu i.e. uv). So, the cycle C contains the edge uv. Now let,

$$S = \{v_i \in C : \text{there is an edge from } u \text{ to } v_{i+1} \text{ in } G\}$$

and

$$T = \{v_j \in C : \text{there is an edge from } v \text{ to } v_j \text{ in } G\}$$

Then, $v_n \notin T$, since otherwise there would be an edge from v to $v_n = v$, i.e., a loop, which is impossible because G is simple graph. Also, $v_n \notin S$(interpreting v_{n+1} as v_1), since otherwise we would again get a loop, this time from u to $v_1 = u$. Thus, $v_n \notin S \cup T$. Let $|S|$, $|T|$ and $|S \cup T|$ denote the number of elements in S, T, and $S \cup T$, respectively. Therefore,

$$|S \cup T| < n \qquad (3.1)$$

Also, for every edge incident with u, there corresponds precisely one vertex v_i in S. Thus,

$$|S| = d(u) \qquad (3.2)$$

Similarly,

$$|T| = d(v) \qquad (3.3)$$

Moreover, if v_k is a vertex belonging to both S and T, there is an edge e joining u to v_{k+1} and an edge f joining v to v_k. This would give

$$C' \equiv v_1 v_{k+1} v_{k+2} \ldots v_n v_k v_{k-1} \ldots v_2 v_1$$

as a Hamiltonian cycle in G, which is a contradiction, since G is non-Hamiltonian (Fig. 3.3). This shows that there is no vertex v_k in $S \cap T$, i.e., $S \cap T = \phi$. Thus, $|S \cup T| = |S| + |T|$. Hence, from Eq. (3.1), (3.2), and (3.3), we have

$$d(u) + d(v) = |S| + |T| = |S \cup T| < n$$

This is impossible. Since, in G, $d(u) \geq {}^n/_2$ and $d(v) \geq {}^n/_2$, and therefore, $d(u) + d(v) \geq n$. This contradiction leads to the conclusion that we have wrongly assumed the result to be false.   □

Theorem 3.4 Let G be a simple graph with n vertices and let u and v be non-adjacent vertices in G such that $d(u) + d(v) \geq n$. Let $G + uv$ denote the supergraph of G obtained by joining u and v by an edge. Then, G is Hamiltonian iff $G + uv$ is Hamiltonian.

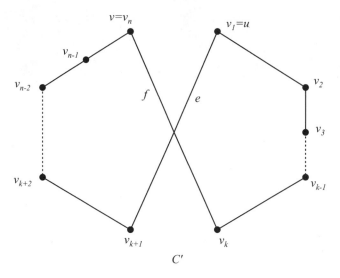

Fig. 3.3 A Hamiltonian cycle C'

Proof Suppose that G is Hamiltonian. Then from the previous Theorem 3.3, the supergraph $G + uv$ must also be Hamiltonian. Conversely, suppose that $G + uv$ is Hamiltonian. Then if G is not Hamiltonian, just as in proof of Theorem 3.3, we can obtain the inequality $d(u) + d(v) < n$. However, by hypothesis, $d(u) + d(v) \geq n$. Hence, G must also be Hamiltonian, as required.   □

The following theorem due to Ore generalizes an earlier result by Dirac (1952).

Ore's Theorem (1962)

Theorem 3.5 *A simple graph with n vertices (where n > 2) is Hamiltonian if the sum of the degrees of every pair of nonadjacent vertices is at least n.*

Proof: Suppose a graph G with n vertices satisfying the given inequality condition is not Hamiltonian. So it is a subgraph of the complete graph K_n with fewer edges. We recursively add edges to the graph by joining nonadjacent vertices until we obtain a graph H such that the addition of one more edge joining two nonadjacent vertices in H will produce a Hamiltonian graph with n vertices. Let x and y be two nonadjacent vertices in H. Thus they are nonadjacent in G also.

Since $d(x) + d(y) \geq n$ in G.

$\Rightarrow d(x) + d(y) \geq n$ in H as well.

If we join the nonadjacent vertices x and y, the resulting graph is Hamiltonian. Hence, in graph H, there is a Hamiltonian path between the vertices x and y. If we write $x = v_1$ and $y = v_n$, this Hamiltonian path can be written as

$$v_1 \rule{1cm}{0.4pt} v_2 \rule{1cm}{0.4pt} \ldots \ v_{i-1} \rule{1cm}{0.4pt} v_i \rule{1cm}{0.4pt} v_{i+1} \rule{1cm}{0.4pt} \ldots \ v_{n-1} \rule{1cm}{0.4pt} v_n$$

Fig. 3.4 A Hamiltonian path from v_1 to v_n

Suppose the degree of v_1 is γ in graph H. If there is an edge between v_1 and v_i in this graph, the existence of an edge between v_{i-1} and v_n will imply that H is Hamiltonian. So whenever vertices v_1 and v_i are adjacent in H, vertices v_n and v_{i-1} are not adjacent (Fig. 3.4). This is true for $1 < i < n$. Hence, $d(v_n) \le (n-1) - \gamma$, since the degree of v_1 is γ. This implies that the sum of the degrees of the two nonadjacent vertices in G is less than n, which contradicts the hypothesis. So any connected graph satisfying the given condition is Hamiltonian (Figs. 3.5 and 3.6).

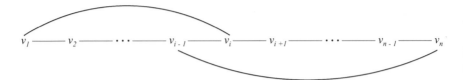

Fig. 3.5 A Hamiltonian cycle $v_1 v_i v_{i+1} \ldots v_n v_{i-1} \ldots v_1$

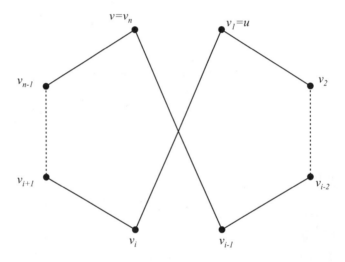

Fig. 3.6 Another representation of Hamiltonian cycle $v_1 v_i v_{i+1} \ldots v_n v_{i-1} \ldots v_1$

Example 3.1 If $K_{m,n}$ be a complete bipartite graph with bipartition (X, Y), then X and Y contains the same number of vertices in $K_{m,n}$, i.e., $|X| = |Y|$
Solution:
Let $K_{m,n}$ has a bipartition (X,Y), where $|X| = m$ and $|Y| = n$. Now, each cycle in $K_{m,n}$ has even length as the graph is bipartite and thus the cycle visits the sets X and Y equally many times, since X and Y are stable subsets. But then necessarily $|X| = |Y| = n$.

Example 3.2 If $G = (V, E)$ is a bipartite graph with bipartition (X,Y), where $|X| = |Y| = n$ and if the degree of each vertex is more than $n/2$, G is Hamiltonian.

$$u = v_1 \text{---} v_2 \; \cdots \; v_{i-1} \text{---} v_i \text{---} v_{i+1} \text{---} \cdots \text{---} v_{2n} = v \text{ in } H$$

Fig. 3.7 A Hamiltonian path from u to v in H

Solution:
Suppose G is not Hamiltonian. Add as many edges as possible joining vertices in X and Y until we obtain a graph H that will become Hamiltonian if one more such edge is added. That is, H is maximal non-Hamiltonian graph. H cannot be complete bipartite $K_{n,n}$. If the degree in G of each vertex is more than $n/2$, the degree of each vertex in H is also more than $n/2$. Let $u \in X$ and $v \in Y$ be two nonadjacent vertices in H. Obviously, there is a Hamiltonian path (Fig. 3.7) from u to v where $v_i \in X$, iff i is odd. If there is an edge joining v_1 and v_i there cannot be an edge joining v_{i-1} and v_{2n}, since H is non-Hamiltonian. Since $d(u) > n/2$, we find that $d(v) < n - \frac{n}{2}$, which contradicts the hypothesis.

[*Note*: If vertex v_i is adjacent to u then v_{i-1} is not adjacent to v. So, if there are r number of such vertices v_i adjacent to u, there must be also r number of vertices v_{i-1} which are not adjacent to v. Now, there are n vertices in X which may be adjacent to v. Therefore,

$$d(v) < n - d(u) \;\Rightarrow\; d(v) < n - \frac{n}{2}, \text{ where } d(u) > n/2.]$$

3.3 Complement and Self-Complementary Graph

Complement: Let G be a simple graph with n vertices. The complement $\bar{G}$ of G is defined to be the simple graph with the same vertex set as G and where two vertices u and v are adjacent precisely when they are not adjacent in G. Intuitively, the complement of G can be obtained from the complete graph K_n by deleting all the edges of G. Figure 3.8 shows a graph G and its complementary graph $\bar{G}$.

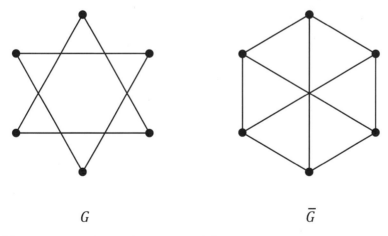

G $\bar{G}$

Fig. 3.8 A graph G and its complementary graph $\bar{G}$

Self-complementary: A simple graph is called self-complementary if it is iso-morphic to its own complement. In Fig. 3.9, the graph G_1 is self-complementary.

Fig. 3.9 A self-complementary graph G_1

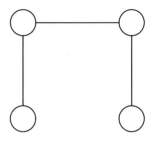

G_1

Example 3.3 Prove that if G is a self-complementary graph with n vertices then n is either $4m$ or $4m + 1$ for some positive integer m.
Solution:
Let $\bar{G}$ be the complement of G. Then $G + \bar{G}$ is a complete graph. The number of edges in complete graph $G + \bar{G}$ is $n(n-1)/2$.

Since G is self-complementary graph G must have $n(n-1)/4$ number of edges.

Therefore, n is congruent to 0 or 1 (modulo 4).

Hence, either $n = 4m$ or $n = 4m + 1$ where m is a positive integer.

Exercises:

1. Find the complements of the following graphs (Fig. 3.10)

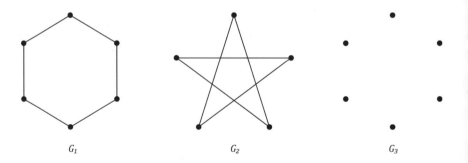

G_1 G_2 G_3

Fig. 3.10

2. Show that the following graphs are self-complementary (Fig. 3.11)

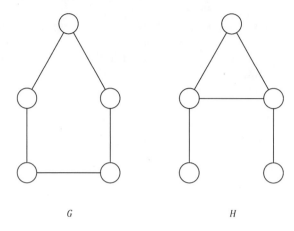

G H

Fig. 3.11

3. Let G be a simple graph with n vertices and let $\bar{G}$ be its complement. Then prove that for each vertex v in G, $d_G(v) + d_{\bar{G}}(v) = n - 1$.

4. Prove that a graph G with n vertices always has a Hamiltonian path if the sum of the degrees of every pair of vertices v_i, v_j in G satisfies the condition

$$d(v_i) + d(v_j) \geq n - 1$$

5. Show that the graph G_1, of Fig. 3.12 is a Hamiltonian and that the graph G_2 has a Hamiltonian path but not a Hamiltonian cycle.

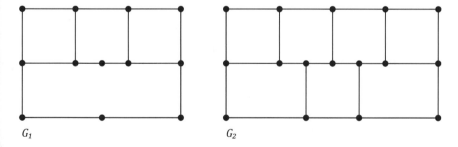

G_1 G_2

Fig. 3.12

6. Let G be a bipartite graph with bipartition $V = X \cup Y$.

   (a) Show that if G is a Hamiltonian then $|X| = |Y|$.

(b) Show that if G is not Hamiltonian but has a Hamiltonian cycle then $|X| = |Y| \pm 1$.

7. Let G be a simple k-regular graph with $2k - 1$ vertices. Prove that G is Hamiltonian.

8. The **Closure** $c(G)$ of a graph G of order n is obtained from G by recursively joining pairs of nonadjacent vertices whose degree sum is at least n until no such pair exists. Show that every graph has a unique closure.

9. Show that a graph is Hamiltonian if and only if its closure is Hamiltonian.

10. Show that a graph is Hamiltonian, if its closure is complete.

11. Give an example of a complete Hamiltonian Bipartite graph.

Chapter 4
Trees and Fundamental Circuits

4.1 Trees

Acyclic graph: A graph with no cycle is acyclic.

Tree: A tree is a connected acyclic graph.

Leaf: A leaf is a vertex of degree 1 (Pendant vertex). A leaf node has no children nodes.

Rooted Tree: The root node of a tree is the node with no parents. There is at most one root node in a rooted tree.

Depth and Level: The depth of a node v is the length of the path from the root to the node v. In a rooted tree, a vertex v is said to be at level $l(v)$ if v is at a distance of $l(v)$ from the root. The set of all nodes at a given depth is sometimes called a *level* of the tree. The root node is at depth zero. Thus the root is at level 0.

Depth or Height of a tree: The depth or height of a tree is the length of the path from the root to the deepest node in the tree. A (rooted) tree with only one node (the root) has a depth of zero.

Forest: A forest is an acyclic graph, which is a collection of trees (Fig. 4.1).

Fig. 4.1 The trees with at most three vertices

Minimally connected graph: A connected graph is said to be minimally connected if the graph becomes disconnected when one edge is removed (Fig. 4.2).

S. Saha Ray, *Graph Theory with Algorithms and its Applications*,
DOI: 10.1007/978-81-322-0750-4_4, © Springer India 2013

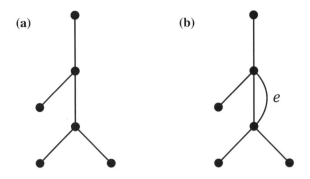

Fig. 4.2 a A minimally connected graph and **b** the graph which is not minimally connected

Binary Tree: A tree in which there is exactly one vertex of degree two and each of the other vertices is of degree one or three is called a *binary tree* (Fig. 4.3).

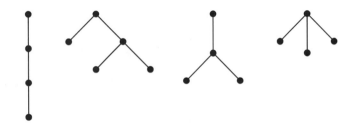

Fig. 4.3 The second tree is only Binary tree among the four trees

Since, the vertex of degree two is distinct from all other vertices, this vertex serves as a *root*. Thus every binary tree is a rooted tree.

Some important properties of binary tree:

1. The number of vertices n in a binary tree is always odd.
2. Let p be the number of pendant vertices in a binary tree T. Then $p = (n + 1)/2$.
3. The number of internal vertices in a binary tree is one less than the number of pendant vertices.

[Hint: Let, q be the number of internal vertices except the root. $1 + p + q = n$ and $2 + p + 3q = 2(n - 1)$. Hence, $p = (n + 1)/2$ and the number of internal vertices $1 + q = n - p = \frac{n+1}{2} - 1$].

4. The minimum height of a n-vertex binary tree is equal to $\lceil \log_2(n+1) - 1 \rceil$
(The ceiling function **ceiling(x)** $= \lceil x \rceil$ is the smallest integer not less than x).

Proof Let l be the height of the binary tree.
Therefore, the maximum level of any vertex of the tree is l.
If n_i denotes the number of vertices at level i, then

$$n_0 = 1, n_1 \leq 2, n_2 \leq 2^2, n_3 \leq 2^3, \ldots, n_l \leq 2^l.$$

Therefore,

$$n = n_0 + n_1 + n_2 + \cdots + n_l \leq 1 + 2 + 2^2 + \cdots + 2^l$$

This implies

$$n \leq 2^{l+1} - 1$$

Consequently,

$$l \geq \log_2(n+1) - 1$$

Hence, the minimum value of $l = \lceil \log_2(n+1) - 1 \rceil$

5. The maximum height of a n-vertex binary tree is equal to $(n-1)/2$.

Proof Let l be the height of the binary tree.
To construct a binary tree with n-vertex such that the farthest vertex is as far as possible from the root, we must have exactly two vertices at each level, except at the root, i.e., at the 0 level.
If n_i denotes the number of vertices at level i, then

$$n_0 = 1, n_1 = 2, n_2 = 2, \ldots, n_l = 2.$$

Therefore,

$$n = n_0 + n_1 + n_2 + \cdots + n_l = 1 + 2l$$

Hence, the maximum value of $l = (n-1)/2$.

4.2 Some Properties of Trees

Theorem 4.1 *Every pair of vertices in a tree is connected by one and only one path.*

Proof Let *T* be a tree; *A, B* be an arbitrary pair of vertices. Since, *T* is a connected graph so *A* and *B* are connected by a path. Let if possible, *A* and *B* be connected by two distinct paths. These two paths together form a circuit and then *T* cannot be a tree. So, there is only one path connecting *A* and *B*.   □

Theorem 4.2 (*Converse of theorem* 4.1) *If there is one and only one path between every pair of vertices in a graph G, then G is a tree.*

Proof Existence of a path between every pair of vertices assures that *G* is connected. Let, if possible, *G* posses a circuit. So, there exists at least one pair of vertices say *A, B* such that there are two distinct paths between *A* and *B*. This contradicts the hypothesis. So *G* does not have any circuit. Hence *G* is a tree. □

Theorem 4.3 *A connected graph is a tree if and only if addition of an edge between any two vertices in the graph creates exactly one circuit.*

Proof If *G* is tree, it is connected and acyclic. If any two non-adjacent vertices are joined by an edge, the unique path in *G* between the two vertices and the edge together form a unique cycle.

Conversely, suppose *G* is connected. There cannot be a cycle in *G* since the supergraph *G'* of *G*, obtained by joining two non-adjacent vertices in *G*, has a unique cycle. So, *G* is a tree.   □

Theorem 4.4 *Any tree with two or more vertices contains at least two pendant vertices.*

Proof Any two vertices in a tree is connected by one and only one path. Since, the tree is supposed to be a finite graph, so there exists a longest path *P*: $v_0 e_0 v_1 e_1 \ldots v_{m-1} e_{m-1} v_m$ in the tree. Since, the tree has at least two vertices so $v_0 \neq v_m$.

Let, if possible, $d(v_0) \neq 1$. Since degree of vertices of a tree with at least two vertices cannot be zero, so $d(v_0) > 1$. So there must be another edge $e \neq e_0$ joining v_0 to a vertex *v* of *T*. If $v = v_i$ for some *i* then the path $v_0 e_0 v_1 e_1 \ldots \ldots v_i e v_o$ forms a circuit. This is impossible, since *T* cannot have any circuit. If *v* is not equal to any v_i of the path *P* then $v e v_0 e_0 v_1 e_1 \ldots v_{m-1} e_{m-1} v_m$ becomes a path of length *m* + 1. This is again a contradiction, since the longest path in *T* has length *m*.

Thus, $d(v_0) = 1$.

Similarly, we can show that $d(v_m) = 1$.

So, we see that v_0 and v_m are pendant and they are distinct. This completes the proof.   □

Theorem 4.5 *A tree with n number of vertices has n − 1 number of edges.*

Proof Let, *T* be a tree. The theorem will be proved by method of induction on *n*. Clearly, the result is true for *n* = 1, 2.

We assume that, the result is true for *k* number of vertices whenever *k* < *n*. In *T* (Fig. 4.4), let *e* be an edge with end vertices *A* and *B*. Since, two vertices in a tree are connected by only one path so there is no other path between *A* and *B*. *e* is the only path joining *A* and *B*. So, *T* − *e* , i.e., the graph obtained from *T* by deleting the

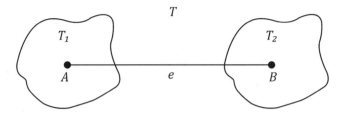

Fig. 4.4 A tree T with minimally connected edge e

edge e becomes a disconnected graph. Now the graph $T - e$ has exactly two components say T_1 and T_2, such that T_1 contains A and T_2 contains B, respectively. Let T_1 and T_2 contain n_1 and n_2 number of vertices. So, $n = n_1 + n_2$. If the component T_1 contains a circuit then T would have a circuit which is not possible. So, T_1 is a tree. Similarly, T_2 is also a tree. So by the hypothesis, T_1 has $n_1 - 1$ and T_2 has $n_2 - 1$ number of edges. Thus, $T - e$ consists of $(n_1 - 1) + (n_2 - 1) = n_1 + n_2 - 2 = n - 2$ number of edges. Hence, T has $n - 2 + 1 = n - 1$ edges. □

Theorem 4.6 (*Converse of theorem* 4.5) *A connected graph with n vertices and n − 1 edges is a tree.*

Proof Let G be a connected graph with n vertices and $n - 1$ edges. Let, if possible, G be not a tree. Then G contains a circuit. Let e be an edge of this circuit. Then the subgraph $G - e$ is still connected. $G - e$ has $n - 2$ edges and n vertices. This is not possible since we know that a connected graph with n vertices has at least $n - 1$ edges. This completes the proof. □

Theorem 4.7 *A graph is a tree if and only if it is minimally connected.*

Proof Let T be a tree having n vertices. So, T has $n - 1$ edges.

If one edge is removed from T then it has $n - 2$ edges. Then, T becomes disconnected since a connected graph with n vertices must have at least $n - 1$ number of edges. Thus T is a minimally connected graph. Conversely, let T be a minimally connected graph with n number of vertices. Since, T is connected graph, so number of edges of $T \geq n - 1$. Let, if possible, T not be a tree. Then, T contains a circuit. T becomes still connected if one edge of this circuit is removed from T. This contradicts our hypothesis that T is a minimally connected graph. Hence T is a tree. □

Theorem 4.8 *A graph with n number of vertices, n − 1 number of edges and without any circuit is connected.*

Proof Let G be a graph. Let, if possible, G be disconnected. Then G has two or more components. Without loss of generality, suppose G_1 and G_2 be two such components. Since, G_1 and G_2 are subgraphs of G they also do not contain any circuit. Let v_j and v_k be two vertices in the components G_1 and G_2 , respectively. Add an edge e between v_j and v_k (Fig. 4.5).

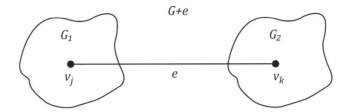

Fig. 4.5 A connected graph $G + e$

Since there is no path between v_j and v_k in G, so adding e would not create a circuit. Thus, the graph G together with e ($i.e. G \cup e$) becomes a connected acyclic graph, i.e., a tree. We see this tree is having n number of vertices and $(n - 1) + 1 = n$ number of edges. This contradicts the fact that a tree having n vertices must have $n - 1$ edges. Hence, G is a connected graph.   □

4.3 Spanning Tree and Co-Tree

A tree T is called a spanning tree of a connected graph G if T is a subgraph of G and if T contains all the vertices of G.

In other words, a spanning tree of a graph G is a spanning subgraph of G that is a tree.

Branch of a tree: An edge of a tree is called a branch of the tree. For example, a, c, f, and h are branches of T, in Fig. 4.6.

Chord of a tree: An edge of G that is not in T is called a chord of T in G. In Fig. 4.6, g is a chord of the spanning tree T in G.

Co-Tree: The complement of a spanning tree T in a connected graph G is called Co-Tree of T. It is denoted by $\bar{T}$. It is illustrated in Fig. 4.6.

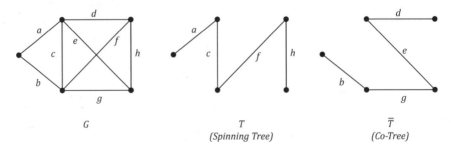

Fig. 4.6 A connected graph G, its spanning tree T, and co-tree $\bar{T}$

4.3.1 Some Theorems on Spanning Tree

Theorem 4.9 *A graph G has a spanning tree iff G is connected.*

Proof Let G be a connected graph. If G has no circuit then it is its own spanning tree. If G contains a circuit then delete an edge from that circuit. Then the graph is still connected. If the graph has no circuit then it becomes a spanning tree of G. Otherwise, repeat the operation till an edge removed from the last circuit yields a connected graph without any circuit. This left out graph contains all vertices of G, because no vertices are removed in the above deletion process. So, this left out tree is a spanning tree of G. Thus, G has a spanning tree.

Conversely, let G be a graph having a spanning tree, say T. Let v_1 and v_2 be two arbitrary vertices of G. Since, T contains all the vertices of G, therefore v_1 and $v_2 \in T$. Since, T is a tree, therefore T is connected and so v_1 and v_2 are connected by a path. So, G is connected.   □

Theorem 4.10 *Let, T be a spanning tree in a connected graph G. G has n vertices and e edges. Then T has $n - 1$ branches and $e - n + 1$ chords.*

Proof This follows from the definition of branch and chord and from the fact that a tree with n vertices contains $n - 1$ edges.

4.4 Fundamental Circuits and Fundamental Cut Sets

4.4.1 Fundamental Circuits

A circuit, formed by adding a chord to a spanning tree of a graph, is called a *fundamental circuit* of the graph with respect to the spanning tree.

The circuit *EGDBCFE* in Fig. 4.7c is a fundamental circuit of the graph 4.7a with respect to 4.7b. This fundamental circuit is obtained by adding the chord *EF* to the spanning tree.

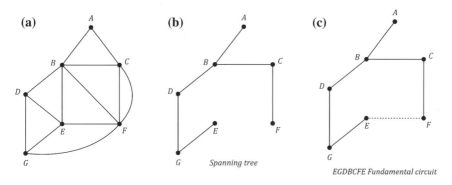

Fig. 4.7 A graph having a spanning tree with corresponding Fundamental circuit *EGDBCFE*

Note: if we add an edge between any two vertices of a tree, a circuit is created. This is because of the fact that there already exists one path between any two vertices of a tree (see Fig. 4.8); adding an edge between them *creates an additional path* and hence a circuit is formed.

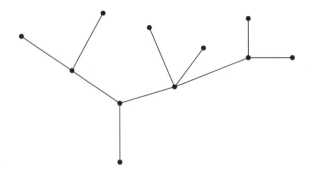

Fig. 4.8 A tree (connected acyclic graph)

Theorem 4.11 *Let T be a spanning tree in a connected graph G. G has n vertices and e edges, then* $\exists e - n + 1$ *number of fundamental circuits formed by T.*

Proof T has $e - n + 1$ number of chords. Hence the theorem follows.

4.4.2 Fundamental Cut Set

Consider a spanning tree T of a connected graph G. In Fig. 4.9, the spanning tree T is represented by the solid lines. Let us take any branch b in T. Since, $\{b\}$ is a cut set in T, $\{b\}$ partitions all vertices of T into two disjoint sets. Consider the same partition of vertices in G and the cut set S in G that corresponds to this partition.

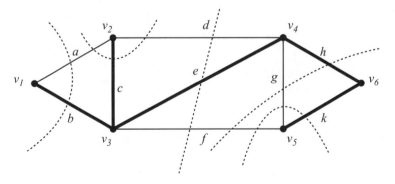

Fig. 4.9 A connected graph showing a spanning tree in solid lines

Cut set S will contain only one branch b of T, and the rest (if any) of the edges in S are chords with respect to T. Such a cut set S containing exactly one branch of a tree T is called a fundamental cut set with respect to T.

Example 4.1 How many fundamental circuits and cut sets are there in a graph G with respect to any spanning tree with 10 vertices and 13 edges.

Solution: Spanning tree with 10 vertices has 9 edges. So, there are 9 fundamental cut sets and there are $13 - 9 = 4$ fundamental circuits.

Example 4.2 Find fundamental circuits for the graph shown below in Fig. 4.10.

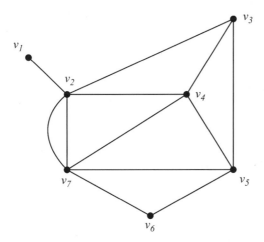

Fig. 4.10

Solution: We have to find a spanning tree and all the corresponding fundamental circuits. First delete all loops and parallels. Consider, the circuit $v_2 - v_3 - v_4 - v_2$. Deleting the edge (v_2, v_3), we get the graph G_1 (Fig. 4.11).

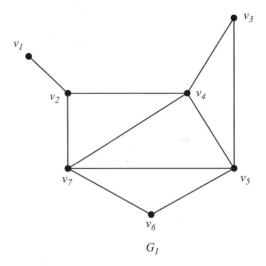

Fig. 4.11

Next, consider the circuit $v_3 - v_4 - v_5 - v_3$. Delete (v_3, v_5) and we get the graph G_2 (Fig. 4.12).

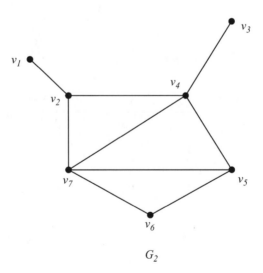

Fig. 4.12

Next, consider the circuit $v_2 - v_4 - v_7 - v_2$. From this delete (v_2, v_4) and we get the graph G_3 (Fig. 4.13).

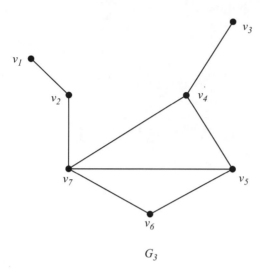

G_3

Fig. 4.13

Next, consider the circuit $v_4 - v_7 - v_5 - v_4$. From this delete (v_4, v_5) and we have the graph G_4 (Fig. 4.14).

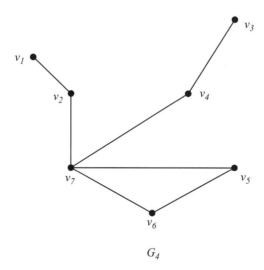

G_4

Fig. 4.14

Finally, delete the edge (v_5, v_6) from the circuit $v_5 - v_6 - v_7 - v_5$ and we get a spanning tree as G_5 (Fig. 4.15)

Fig. 4.15 A spanning tree G_5

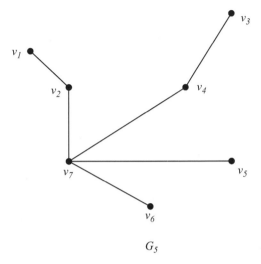

G_5

Now, the number of edges of the given graph, after converting it into a simple one, $e = 11$, number of vertices $n = 7$. So there are $e - n + 1 = 11 - 7 + 1 = 5$ fundamental circuits which are $v_2 - v_4 - v_7 - v_2, v_2 - v_3 - v_4 - v_7 - v_2, v_4 - v_5 - v_7 - v_4, v_5 - v_6 - v_7 - v_5, v_3 - v_4 - v_7 - v_5 - v_3$.

Exercises:

1. Find the spanning tree of the following graph in Fig. 4.16. Hence find out the fundamental circuits and fundamental cut sets.

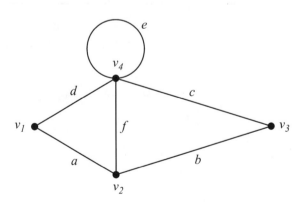

Fig. 4.16

2. Prove that any tree with at least two vertices is a bipartite graph.
3. Let T be a tree and let u and v be two vertices of T which are not adjacent. Let G be the supergraph of T obtained from T by joining u and v by an edge. Prove that G contains a cycle.
4. Let T be a tree with n vertices, where $n \geq 4$, and let v be a vertex of maximum degree in T.
   (a) Show that T is a path if and only if $d(v) = 2$.
   (b) Prove that if $d(v) = n - 2$ then any other tree with n vertices and maximum vertex degree $n - 2$ is isomorphic to T.
   (c) Prove that if $n \geq 6$ and $d(v) = n - 3$ then there are exactly 3 non-isomorphic trees which T can be.
5. Let T be a tree with n vertices, where $n \geq 3$. Show that there is a vertex v in T with $d(v) \geq 2$ such that every vertex adjacent to v, except possibly for one, has degree 1.
6. Let T be a tree and let v be a vertex of maximum degree in T, say $d(v) = k$. Prove that T has at least k vertices of degree 1.
7. Let T be a tree with at least k edges, $k \geq 2$. How many connected components are there in the subgraph of T obtained by deleting k edges of T ?
8. Let G be a connected graph which is not a tree and let C be a cycle in G. Prove that the complement of any spanning tree of G contains at least one edge of C.
9. Let G be a graph with exactly one spanning tree. Prove that G is a tree.
10. An edge e (not a loop) of a graph G is said to be *contracted* if it is deleted and then its end vertices are fused. The resulting graph is denoted by $G * e$ illustrated in Fig. 4.17.

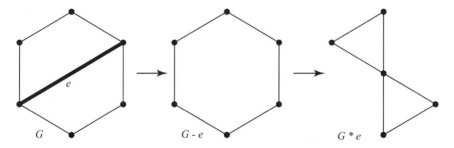

Fig. 4.17 A graph G showing edge deleted subgraph $G - e$ and contracted subgraph $G * e$

   (a) Prove that if T is a spanning tree of G which contains e then $T * e$ is a spanning tree of $G * e$.
   (b) Prove that if T' is a spanning tree of $G * e$ then there is a unique spanning tree T of G which contains e and is such that $T' = T * e$.
11. Show that a Hamiltonian path is a spanning tree.

Chapter 5
Algorithms on Graphs

5.1 Shortest Path Algorithms

Weighted network: A weighted network (V, E, C) consists of a node set V, an edge set E, and the weight set C specifying weights c_{ij} for the edges $(i, j) \in E$.

5.1.1 Dijkstra's Algorithm

We determine the shortest route or shortest distance along with the shortest path between any two fixed pair of vertices of a directed or undirected graph.

One of the most important and useful algorithm is *Dijkstra's shortest path algorithm*, a *greedy* algorithm that efficiently finds shortest paths in a graph. A greedy algorithm for an optimization problem always makes the choice that looks best at the moment and adds it to the current subsolution. It builds a solution by repeatedly selecting the locally optimal choice among all options at each stage.

Because graphs are able to represent many things, many problems can be cast as shortest-path problems, making Dijkstra's algorithm a powerful and general tool.

5.1.1.1 Applications of Dijkstra's Algorithm

- Dijkstra's algorithm is applied to automatically find directions between physical locations, such as driving directions on websites like Mapquest or Google Maps.
- In a networking or telecommunication applications, Dijkstra's algorithm has been used for solving the min-delay path problem (which is the shortest path problem). For example in data network routing, the goal is to find the path for data packets to go through a switching network with minimal delay.

S. Saha Ray, *Graph Theory with Algorithms and its Applications*,
DOI: 10.1007/978-81-322-0750-4_5, © Springer India 2013

• It is also used for solving a variety of shortest path problems arising in plant and facility layout, robotics, transportation, and very large-scale integration (VLSI) design.

Dijkstra's algorithm solves shortest path problem to find the shortest path from a given node s, called a starting node or an initial node, to all other nodes in the network. This algorithm solves only the problems with non-negative costs (weights), i.e., $c_{ij} \geq 0$ for all $(i,j) \in E$.

The algorithm characterizes each node by its state. The state of a node consists of two features: *distance value* and *status label*.

• Distance value of a node is a scalar representing an estimate of its distance from node s.
• Status label is an attribute specifying whether the distance value of a node is equal to the shortest distance to node s or not.

(a) The status label of a node is *Permanent* if its distance value is equal to the shortest distance from node s
(b) Otherwise, the status label of a node is *Temporary*

The algorithm maintains and step-by-step updates the states of the nodes.
At each step one node is designated as *current*.

5.1.1.2 Notations for Dijkstra's Algorithm

• *Label of a vertex*: The label of a vertex v is defined as length of the shortest distance from the source vertex s to the corresponding vertex v and it is denoted by $l(v)$.
• P or T denotes the status label of a node, where P stands for *permanent* and T stands for *temporary*.
• c_{ij} is the cost (weight $w(i,j)$ or w_{ij}) of traversing link (i,j) as given by the problem

The state of a node v is the ordered pair of its distance value $l(v)$ and its status label.

Dijkstra's Algorithm:

Algorithm 5.1 *Step* 1: Set $l(s) = 0$ and mark the label of s as permanent. And for all the vertices $v \neq s$ (non starting vertices), assign a temporary label $l(v) = \infty$ [The state of node s is $(0, P)$ and the state of every other node is (∞, T)] and set $u = s$ (Designate the node s as the *current* node).

Step 2: If u be a vertex with permanent label, then for every edge $e = uv$ incident with u, if $l(v) > l(u) + w(e)$ (where $w(e)$ is the weight of the edge $e(= uv)$) and v is the temporary label vertex then [Update the label (distance values) of these nodes]
set $l(v) = l(u) + w(e)$ and predecessor$(v) = u$.

Step 3: Let k be a temporary label vertex for which $l(k)$ is minimum. If no such k vertex exists then ∃ no shortest path from source vertex s to destination vertex t (say). Otherwise go to step 4.

Step 4: Make the label of k as permanent. If $k = t$ then stop. Otherwise set $u = k$ [Designate this node as the *current* node] then go to step 2. □

Dijkstra's algorithm starts by assigning some initial values for the distances from node s and to every other node in the network. It operates in steps, where at each step the algorithm improves the distance values. At each step, the shortest distance from node s to another node is determined.

5.1.1.3 Complexity

Dijkstra's algorithm solves shortest path problem in $O(|V|^2)$ time. The algorithm contains an outer loop executed $|V| - 1$ times and inner loops, to find the closest vertex and update distances, executed $O(|V|)$ times for each iteration of the outer loop. Its time-complexity is therefore $O(|V|^2)$, i.e., $O(n^2)$, where $|V| = n$.

Example 5.1 Apply Dijskra's Algorithm, to find the shortest route from node 1 to 6 of Fig. 5.1.

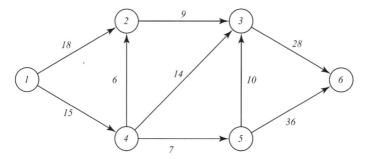

Fig. 5.1

Solution:
Initially, assign label 0 to the starting node 1 and make it permanent. Other non-starting nodes must be assigned label ∞ and temporary status (Table 5.1).

Table 5.1

Vertex (v)	1	2	3	4	5	6
Label (v)	0	∞	∞	∞	∞	∞
Status (v)	P	T	T	T	T	T
Predecessor (v)	–	–	–	–	–	–

The temporary label vertices 2 and 4 are adjacent to 1. In Table 5.2, we update the label of these vertices and also the predecessor of them must be 1.

Table 5.2 Adjacent vertices of 1 are 2 and 4

Vertex (v)	1	2	3	4	5	6
Label (v)	0	18	∞	15	∞	∞
Status (v)	P	T	T	T	T	T
Predecessor (v)	–	1	–	1	–	–

We search the minimum among temporary labeled vertices. Min $(18, \infty, 15, \infty, \infty) = 15$ for which temporary labeled vertex 4 becomes permanent in Table 5.3.

Table 5.3 Minimum temporary labeled vertex 4 becomes permanent

Vertex (v)	1	2	3	**4**	5	6
Label (v)	0	18	∞	**15**	∞	∞
Status (v)	P	T	T	**P**	T	T
Predecessor (v)	–	1	–	**1**	–	–

The temporary labeled vertices adjacent to vertex 4 are 2, 3, and 5. For the vertex 2, $l(4) + w(4, 2) = 15 + 6 = 21$, but in Table 5.3, label of node 2 was 18. So, no change is required. In Table 5.4, we update the label of vertices 3 and 5 and also the predecessor of them must be 4.

Table 5.4 Adjacent vertices of 4 are 2, 3, and 5

Vertex (v)	1	**2**	**3**	4	**5**	6
Label (v)	0	18	29	15	22	∞
Status (v)	P	T	T	P	T	T
Predecessor (v)	–	1	4	1	4	–

Again, we search the minimum among temporary labeled vertices. Min $(18, 29, 22, \infty) = 18$ for which temporary labeled vertex 2 becomes permanent in Table 5.5.

Table 5.5 Minimum temporary labeled vertex 2 becomes permanent

Vertex (v)	1	**2**	3	4	5	6
Label (v)	0	**18**	29	15	22	∞
Status (v)	P	**P**	T	P	T	T
Predecessor (v)	–	**1**	4	1	4	–

Only vertex 3 is adjacent to 2. Since, $l(2) + w(2, 3) = 27$. In Table 5.6, the new label of 3 will be 27 and the corresponding predecessor will become 2.

Table 5.6 Adjacent vertex of 2 is only 3

Vertex (v)	1	2	**3**	4	5	6
Label (v)	0	18	27	15	22	∞
Status (v)	P	P	T	P	T	T
Predecessor (v)	–	1	2	1	4	–

Since, the minimum among temporary labeled vertex is 5, it becomes permanent in Table 5.7.

Table 5.7 Minimum temporary labeled vertex 5 becomes permanent

Vertex (v)	1	2	3	4	5	6
Label (v)	0	18	27	15	**22**	∞
Status (v)	P	P	T	P	**P**	T
Predecessor (v)	–	1	2	1	**4**	–

The temporary labeled vertices adjacent to vertex 5 are 3 and 6.
In Table 5.8,

Table 5.8 Adjacent vertices of 5 are 3 and 6

Vertex (v)	1	2	**3**	4	5	**6**
Label (v)	0	18	27	15	22	58
Status (v)	P	P	T	P	P	T
Predecessor (v)	–	1	2	1	4	5

1. For the vertex 3, $l(5) + w(5,3) = 22 + 10 = 32$. But in Table 5.7, label of vertex 3 was 27. Consequently no change is required, since min $(27, 32) = 27$.
2. For the vertex 6, we update the new label of 6 and also update the predecessor of it accordingly.

   In Table 5.9, here, min $(27, 58) = 27$. So, label of 3 has been made permanent.

Table 5.9 Minimum temporary labeled vertex 3 becomes permanent

Vertex (v)	1	2	**3**	4	5	6
Label (v)	0	18	**27**	15	22	58
Status (v)	P	P	**P**	P	P	T
Predecessor (v)	–	1	**2**	1	4	5

Only vertex 6 is adjacent to 3. In Table 5.10, for the vertex 6, $l(3) + w(3,6) = 27 + 28 = 55$, but from the Table 5.9, min $(55, 58) = 55$. Therefore, the new label of 6 is 55 and the updated predecessor will be 3.

Table 5.10 Adjacent vertex of 3 is 6

Vertex (v)	1	2	3	4	5	**6**
Label (v)	0	18	27	15	22	55
Status (v)	P	P	P	P	P	T
Predecessor (v)	–	1	2	1	4	3

Table 5.11 The destination vertex 6 becomes permanent

Vertex (v)	1	2	3	4	5	**6**
Label (v)	0	18	27	15	22	**55**
Status (v)	P	P	P	P	P	**P**
Predecessor (v)	–	1	2	1	4	**3**

Since, the destination vertex 6 becomes permanent, we shall stop here.

The required shortest distance from node 1 to 6 is 55 units, which is the permanent label of the destination node.

To determine the shortest path, we backtrack from the destination node 6 to starting node 1. From Tables 5.11, 5.9, and 5.5, we see that the predecessor of 6 is 3, predecessor of 3 is 2 and predecessor of 2 is 1 respectively.

Hence, the required shortest path is 1–2–3–6.

Moreover, it can be verified from the network in Fig. 5.1, the sum of the weights of edges along the shortest path is 55.

Alternative approach:

Following the same argument as discussed above the Table 5.12 can be constructed.

Table 5.12

Vertex (v)	1	2	3	4	5	6
	$\boxed{0}$	∞	∞	∞	∞	∞
	0	18	∞	$\boxed{15}$	∞	∞
	0	$\boxed{18}$	29	15	22	∞
	0	18	27	15	$\boxed{22}$	∞
	0	18	$\boxed{27}$	15	22	58
	0	18	27	15	22	$\boxed{55}$

In Table 5.12, the permanent labels of the vertices are enclosed by the squares.

' Since, the destination vertex 6 becomes permanent, we shall stop here.

The required shortest distance from vertex 1 to 6 is 55 units, which is the permanent label of the destination vertex.

To determine the shortest path, we backtrack from the destination vertex 6 to starting vertex 1. From Table 5.12, we see that label of vertex 6 changes to 55 from that row in which the permanent labeled vertex is 3. So, the predecessor of 6 is 3.

Again, label of vertex 3 changes to 27 from that row in which the permanent labeled vertex is 2. Therefore, the predecessor of 3 is 2.

According to similar argument, the predecessor of 2 is 1.

Hence, the required shortest path is 1–2–3–6.

Example 5.2

A truck must deliver concrete from the ready-mix plant to a construction site. The network in Fig. 5.2 represents the available routes between the plant and the site. The distances from node-to-node are given along the route lines. Use *Dijkstra's Algorithm*, to determine the best route from plant to site.

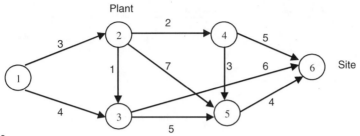

Fig. 5.2

Solution:
Initially, assign label 0 to the starting vertex 2 and make it permanent. Other non-starting vertices must be assigned label ∞ and temporary status (Table 5.13).

Table 5.13

Vertex (v)	1	2	3	4	5	6
Label (v)	∞	**0**	∞	∞	∞	∞
Status (v)	T	**P**	T	T	T	T
Predecessor (v)	–	–	–	–	–	–

The temporary label vertices 3, 4, and 5 are adjacent to 2. We update the label of these vertices and also the predecessor of them must be 2 (Table 5.14).

Table 5.14 Adjacent vertices of 2 are 3, 4, 5

Vertex (v)	1	2	**3**	**4**	**5**	6
Label (v)	∞	0	1	2	7	∞
Status (v)	T	P	T	T	T	T
Predecessor (v)	–	–	2	2	2	–

We search the minimum among temporary labeled vertices. Min $(\infty, 1, 2, 7, \infty) = 1$ for which temporary labeled vertex 3 becomes permanent in Table 5.15.

Table 5.15 Minimum temporary labeled vertex 3 becomes permanent

Vertex (v)	1	2	**3**	4	5	6
Label (v)	∞	0	**1**	2	7	∞
Status (v)	T	P	**P**	T	T	T
Predecessor (v)	–	–	**2**	2	2	–

The temporary label vertices 5 and 6 are adjacent to 3.
For the vertex 5, $l(3) + w(3,5) = 1 + 5 = 6$, but in Table 5.14, label of node 5 was 7. Since, min $(7, 6) = 6$, the new label of vertex 5 will be 6 in Table 5.16.
For the vertex 6, $l(3) + w(3,6) = 1 + 6 = 7$, and min $(\infty, 7) = 7$.
We update the label of vertex 6 and the predecessor of them must be 3 in Table 5.16.

Table 5.16 Adjacent vertices of 3 are 5 and 6

Vertex (v)	1	2	3	4	**5**	**6**
Label (v)	∞	0	1	2	6	7
Status (v)	T	P	P	T	T	T
Predecessor (v)	–	–	2	2	3	3

Again, we search the minimum among temporary labeled vertices. Min $(\infty, 2, 6, 7) = 2$ for which temporary labeled vertex 4 becomes permanent (Table 5.17).

Table 5.17 Minimum temporary labeled vertex 4 becomes permanent

Vertex (v)	1	2	3	**4**	5	6
Label (v)	∞	0	1	**2**	6	7
Status (v)	T	P	P	**P**	T	T
Predecessor (v)	–	–	2	**2**	3	3

The vertices 5 and 6 are adjacent to 4.

For the vertex 5, $l(4) + w(4, 5) = 2 + 3 = 5$, but in Table 5.17, label of node 5 was 6. Since, min $(6, 5) = 5$, the new label of vertex 5 will be 5.

For the vertex 6, $l(4) + w(4, 6) = 2 + 5 = 7$, and so, no change is required.

We update the label of vertex 5 and the predecessor of it must be 4 (Table 5.18).

Table 5.18 Adjacent vertices of 4 are 5 and 6

Vertex (v)	1	2	3	4	**5**	**6**
Label (v)	∞	0	1	2	5	7
Status (v)	T	P	P	P	T	T
Predecessor (v)	–	–	2	2	4	3

Again, we search the minimum among temporary labeled vertices. Min $(\infty, 5, 7) = 5$ for which temporary labeled vertex 5 becomes permanent in Table 5.19.

Table 5.19 Minimum temporary labeled vertex 5 becomes permanent

Vertex (v)	1	2	3	4	**5**	6
Label (v)	∞	0	1	2	**5**	7
Status (v)	T	P	P	P	**P**	T
Predecessor (v)	–	–	2	2	**4**	3

Only vertex 6 is adjacent to 5. Since, $l(5) + w(5, 6) = 5 + 4 = 9$. The label of 6 will remain unchanged. Consequently, Table 5.19 will be unaltered.

Now, we search the minimum among temporary labeled vertices. Min $(\infty, 7) = 7$ for which temporary labeled vertex 6 becomes permanent in Table 5.20.

Table 5.20 Minimum temporary labeled vertex 6 becomes permanent

Vertex (v)	1	2	3	4	5	6
Label (v)	∞	0	1	2	5	7
Status (v)	T	P	P	P	P	P
Predecessor (v)	–	–	2	2	4	3

Since, the destination vertex 6 becomes permanent, we shall stop here.

The required shortest distance from vertex (Plant) 2 to (Site) 6 is 7 units, which is the permanent label of the destination vertex.

To determine the shortest path, we backtrack from the destination vertex 6 to starting vertex 2. From Tables 5.20, and 5.15, we see that the predecessor of 6 is 3, and predecessor of 3 is 2 respectively.

Hence, the required shortest path is 2–3–6.

Moreover, it can be verified from the network in Fig. 5.2, the sum of the weights of edges along the shortest path is 7.

Alternative approach:
Following the same argument as discussed above the following table can be constructed. In Table 5.21, the permanent labels of the vertices are enclosed by the squares.

Table 5.21

Vertex (v)	1	2	3	4	5	6
	∞	[0]	∞	∞	∞	∞
	∞	0	[1]	2	7	∞
	∞	0	1	[2]	6	7
	∞	0	1	2	[5]	7
	∞	0	1	2	5	[7]

Since, the destination vertex 6 becomes permanent, we shall stop here.

The required shortest distance from vertex 2 to 6 is 7 units, which is the permanent label of the destination vertex.

To determine the shortest path, we backtrack from the destination vertex 6 to starting vertex 2. From Table 5.21, we see that label of vertex 6 changes to 7 from that row in which the permanent labeled vertex is 3. So, predecessor of 6 is 3.

Again, label of vertex 3 changes to 1 from that row in which the permanent labeled vertex is 2. So, predecessor of 3 is 2.

Hence, the required shortest path is 2–3–6.

5.1.2 Floyd-Warshall's Algorithm

The problem of finding the shortest path between all pairs of vertices on a graph is akin to making a table of all of the distances between all pairs of cities on a road map.

The Floyd-Warshall All-Pairs-Shortest-Path algorithm uses a dynamic-programming methodology to solve the All-Pairs-Shortest-Path problem. It uses a recursive approach to find the minimum distances between all nodes in a graph. The striking feature of this algorithm is its usage of dynamic programming to avoid redundancy and thus solving the All-Pairs-Shortest-Path problem in $O(n^3)$.

This algorithm is more general than Dijkstra's Algorithm because it determines the shortest route between any two nodes or vertices in the network.

5.1.2.1 Applications of Floyd-Warshall's Algorithm

The Floyd-Warshall's Algorithm can be used to solve the following problems, among others:

1. Shortest paths in directed graph.
2. Transitive closure of directed graphs.
3. Finding a regular expression denoting the regular language accepted by a finite automaton (Kleen's Algorithm).
4. Inversion of real matrices (Gauss Jardon Algorithm).
5. Optimal routing. In this application one is interested in finding the path with the maximal flow between two vertices.
6. Testing whether an undirected graph is bipartite.

To determine the shortest route between every pair of vertices in the network, this algorithm requires two matrices, viz.

1. Distance Matrix D

   and

2. Node Sequence Matrix S

   Floyd-Warshall's Algorithm:

Algorithm 5.2 *Initial step*: Define the initial distance matrix $D^{(0)}$ and initial node sequence matrix $S^{(0)}$ as follows:

1. Set all diagonal elements of $D^{(0)}$ to 0.

   That is $d_{ii}^{(0)} = 0, \quad i = 1, 2, \ldots, n$

2. And also, set all elements of $S^{(0)}$ to zero. i.e., $s_{ij}^{(0)} = 0 \quad \forall i, j = 1, 2, \ldots, n$.

   Then set $k = 1$.
   Step-k: $(1 \leq k \leq n)$
   If the condition $d_{ij}^{(k-1)} > d_{ik}^{(k-1)} + d_{kj}^{(k-1)}$ (for $i, j = 1, 2, \ldots, n$. Provided $i \neq k$; $j \neq k$; $i \neq j$) is satisfied (Fig. 5.3)

1. Create $D^{(k)}$ by replacing $d_{ij}^{(k-1)}$ in $D^{(k-1)}$ with $d_{ik}^{(k-1)} + d_{kj}^{(k-1)}$.
2. Create node sequence matrix $S^{(k)}$ by replacing $s_{ij}^{(k-1)}$ in $S^{(k-1)}$ with k.

   Then, set $k = k + 1$, repeat step k until $k = n$.

After n steps, we can determine the shortest route between vertices i and j from the two matrices $D^{(n)}$ and $S^{(n)}$ using the following rules:

1. From $D^{(n)}$, $d_{ij}^{(n)}$ gives the shortest distance between vertices i and j.
2. From $S^{(n)}$, determine the intermediate vertex $k = s_{ij}^{(n)}$ which yields the route $i - k - j$. If $k = 0$, then stop. Otherwise, repeat the procedure between vertices i and k and vertices k and j. $\qquad\square$

For a path $P \equiv v_1 - v_2 \ldots - v_l$, we say that the vertices $v_2, v_3, \ldots, v_{l-1}$ are the *intermediate vertices* of this path. Note that a path consisting of a single edge has no intermediate vertices. We define $d_{ij}^{(k)}$ to be the distance along the shortest path from i to j such that any intermediate vertices on the path are chosen from the set $\{1, 2, \ldots, n\}$, as cited in Fig. 5.3. In other words, we consider a path from i to j which either consists of the single edge (i, j), or it visits some intermediate vertices along the way, but these intermediate can only be chosen from $\{1, 2, \ldots, n\}$. The path is free to visit any subset of these vertices, and to do so in any order.

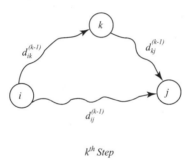

k^{th} Step

Fig. 5.3

5.1.2.2 Complexity

With three nested loops of general Step k, Floyd's algorithm runs in $O(|V|^3)$-time. Therefore, the time Complexity of this algorithm is $O(n^3)$, where $|V| = n$.

Example 5.3 Applying *Floyd-Warshall's Algorithm*, determine the shortest route from node 1 to 4 and from node 2 to 4 of the following Network (Fig. 5.4).

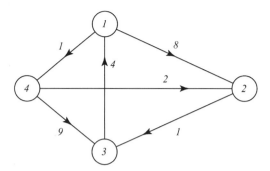

Fig. 5.4

Solution:
Initial step:

$$
D^{(0)} = \begin{array}{c} \\ 1 \\ 2 \\ 3 \\ 4 \end{array}
\begin{array}{cccc} 1 & 2 & 3 & 4 \end{array}
\left[\begin{array}{cccc}
0 & 8 & \infty & 1 \\
\infty & 0 & 1 & \infty \\
4 & \infty & 0 & \infty \\
\infty & 2 & 9 & 0
\end{array} \right],
\quad
S^{(0)} = \begin{array}{c} \\ 1 \\ 2 \\ 3 \\ 4 \end{array}
\begin{array}{cccc} 1 & 2 & 3 & 4 \end{array}
\left[\begin{array}{cccc}
0 & 0 & 0 & 0 \\
0 & 0 & 0 & 0 \\
0 & 0 & 0 & 0 \\
0 & 0 & 0 & 0
\end{array} \right]
$$

At each Step $k(k = 1, 2\ldots, n = 4)$, node k is considered as intermediate node between every pair of nodes. The kth row, kth column, and diagonal elements of $D^{(k-1)}$ and $S^{(k-1)}$ will remain unchanged in Step k.

Step-1:

$$
D^{(1)} = \begin{array}{c} \\ 1 \\ 2 \\ 3 \\ 4 \end{array}
\begin{array}{cccc} 1 & 2 & 3 & 4 \end{array}
\left[\begin{array}{cccc}
0 & 8 & \infty & 1 \\
\infty & 0 & 1 & \infty \\
4 & 12 & 0 & 5 \\
\infty & 2 & 9 & 0
\end{array} \right],
\quad
S^{(1)} = \begin{array}{c} \\ 1 \\ 2 \\ 3 \\ 4 \end{array}
\begin{array}{cccc} 1 & 2 & 3 & 4 \end{array}
\left[\begin{array}{cccc}
0 & 0 & 0 & 0 \\
0 & 0 & 0 & 0 \\
0 & 1 & 0 & 1 \\
0 & 0 & 0 & 0
\end{array} \right]
$$

Step-2:

$$
D^{(2)} = \begin{array}{c} \\ 1 \\ 2 \\ 3 \\ 4 \end{array}
\begin{array}{cccc} 1 & 2 & 3 & 4 \end{array}
\left[\begin{array}{cccc}
0 & 8 & 9 & 1 \\
\infty & 0 & 1 & \infty \\
4 & 12 & 0 & 5 \\
\infty & 2 & 3 & 0
\end{array} \right],
\quad
S^{(2)} = \begin{array}{c} \\ 1 \\ 2 \\ 3 \\ 4 \end{array}
\begin{array}{cccc} 1 & 2 & 3 & 4 \end{array}
\left[\begin{array}{cccc}
0 & 0 & 2 & 0 \\
0 & 0 & 0 & 0 \\
0 & 1 & 0 & 1 \\
0 & 0 & 2 & 0
\end{array} \right]
$$

Step-3:

$$
D^{(3)} = \begin{array}{c} \\ 1 \\ 2 \\ 3 \\ 4 \end{array}
\begin{array}{c} \begin{array}{cccc} 1 & 2 & 3 & 4 \end{array} \\
\begin{bmatrix} 0 & 8 & 9 & 1 \\ 5 & 0 & 1 & 6 \\ 4 & 12 & 0 & 5 \\ 7 & 2 & 3 & 0 \end{bmatrix} \end{array}, \quad
S^{(3)} = \begin{array}{c} \\ 1 \\ 2 \\ 3 \\ 4 \end{array}
\begin{array}{c} \begin{array}{cccc} 1 & 2 & 3 & 4 \end{array} \\
\begin{bmatrix} 0 & 0 & 2 & 0 \\ 3 & 0 & 0 & 3 \\ 0 & 1 & 0 & 1 \\ 3 & 0 & 2 & 0 \end{bmatrix} \end{array}
$$

Step-4:

$$
D^{(4)} = \begin{array}{c} \\ 1 \\ 2 \\ 3 \\ 4 \end{array}
\begin{array}{c} \begin{array}{cccc} 1 & 2 & 3 & 4 \end{array} \\
\begin{bmatrix} 0 & 3 & 4 & 1 \\ 5 & 0 & 1 & 6 \\ 4 & 7 & 0 & 5 \\ 7 & 2 & 3 & 0 \end{bmatrix} \end{array}, \quad
S^{(4)} = \begin{array}{c} \\ 1 \\ 2 \\ 3 \\ 4 \end{array}
\begin{array}{c} \begin{array}{cccc} 1 & 2 & 3 & 4 \end{array} \\
\begin{bmatrix} 0 & 4 & 4 & 0 \\ 3 & 0 & 0 & 3 \\ 0 & 4 & 0 & 1 \\ 3 & 0 & 2 & 0 \end{bmatrix} \end{array}
$$

Now, from matrices $D^{(4)}$ and $S^{(4)}$, we can now determine the shortest route between every pair of nodes.

The shortest distance from node 1 to node 4 is $d_{14}^{(4)} = 1$ unit.

To determine the associated path, we determine the intermediate node $k = s_{14}^{(4)} = 0$, which indicates that there is no intermediate node between 1 and 4. Therefore, the shortest path is 1–4.

Again, the shortest distance from node 2 to node 4 is $d_{24}^{(4)} = 6$ units. To determine the associated path, we determine the intermediate node $k = s_{24}^{(4)} = 3$ between nodes 2 and 4, yields the route 2–3–4. Since, $s_{23}^{(4)} = 0$, no further intermediate node exists between nodes 2 and 3. From $S^{(4)}$, we determine the intermediate node $s_{34}^{(4)} = 1$ between nodes 3 and 4, yields the route 3–1–4. But, since, $s_{31}^{(4)} = 0$ and $s_{14}^{(4)} = 0$, so no further node exists between nodes 3 and 1 and also between nodes 1 and 4, respectively.

The combined result now gives the shortest path as 2–3–1–4. The associated length of the route is 6 units.

Example 5.4 Using *Floyd-Warshall's Algorithm*, determine the shortest route from node 1 to 5 of the following Network (Fig. 5.5).

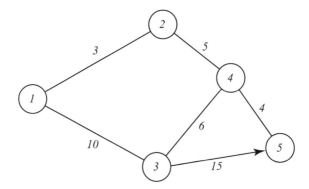

Fig. 5.5

Solution:
Initial step:

$$
D^{(0)} = \begin{array}{c} \\ 1 \\ 2 \\ 3 \\ 4 \\ 5 \end{array}
\begin{array}{ccccc}
1 & 2 & 3 & 4 & 5 \\
\left[\begin{array}{ccccc}
0 & 3 & 10 & \infty & \infty \\
3 & 0 & \infty & 5 & \infty \\
10 & \infty & 0 & 6 & 15 \\
\infty & 5 & 6 & 0 & 4 \\
\infty & \infty & \infty & 4 & 0
\end{array}\right]
\end{array},
\quad
S^{(0)} = \begin{array}{c} \\ 1 \\ 2 \\ 3 \\ 4 \\ 5 \end{array}
\begin{array}{ccccc}
1 & 2 & 3 & 4 & 5 \\
\left[\begin{array}{ccccc}
0 & 0 & 0 & 0 & 0 \\
0 & 0 & 0 & 0 & 0 \\
0 & 0 & 0 & 0 & 0 \\
0 & 0 & 0 & 0 & 0 \\
0 & 0 & 0 & 0 & 0
\end{array}\right]
\end{array}
$$

At each Step $k(k = 1, 2\ldots, n = 5)$, node k is considered as intermediate node between every pair of nodes. The kth row, kth column, and diagonal elements of $D^{(k-1)}$ and $S^{(k-1)}$ will remain unchanged in Step k.

Step-1:

$$
D^{(1)} = \begin{array}{c} \\ 1 \\ 2 \\ 3 \\ 4 \\ 5 \end{array}
\begin{array}{ccccc}
1 & 2 & 3 & 4 & 5 \\
\left[\begin{array}{ccccc}
0 & 3 & 10 & \infty & \infty \\
3 & 0 & 13 & 5 & \infty \\
10 & 13 & 0 & 6 & 15 \\
\infty & 5 & 6 & 0 & 4 \\
\infty & \infty & \infty & 4 & 0
\end{array}\right]
\end{array},
\quad
S^{(1)} = \begin{array}{c} \\ 1 \\ 2 \\ 3 \\ 4 \\ 5 \end{array}
\begin{array}{ccccc}
1 & 2 & 3 & 4 & 5 \\
\left[\begin{array}{ccccc}
0 & 0 & 0 & 0 & 0 \\
0 & 0 & 1 & 0 & 0 \\
0 & 1 & 0 & 0 & 0 \\
0 & 0 & 0 & 0 & 0 \\
0 & 0 & 0 & 0 & 0
\end{array}\right]
\end{array}
$$

Step-2:

$$
D^{(2)} = \begin{array}{c} \\ 1 \\ 2 \\ 3 \\ 4 \\ 5 \end{array}
\begin{array}{ccccc}
1 & 2 & 3 & 4 & 5 \\
\left[\begin{array}{ccccc}
0 & 3 & 10 & 8 & \infty \\
3 & 0 & 13 & 5 & \infty \\
10 & 13 & 0 & 6 & 15 \\
8 & 5 & 6 & 0 & 4 \\
\infty & \infty & \infty & 4 & 0
\end{array}\right]
\end{array},
\quad
S^{(2)} = \begin{array}{c} \\ 1 \\ 2 \\ 3 \\ 4 \\ 5 \end{array}
\begin{array}{ccccc}
1 & 2 & 3 & 4 & 5 \\
\left[\begin{array}{ccccc}
0 & 0 & 0 & 2 & 0 \\
0 & 0 & 1 & 0 & 0 \\
0 & 1 & 0 & 0 & 0 \\
2 & 0 & 0 & 0 & 0 \\
0 & 0 & 0 & 0 & 0
\end{array}\right]
\end{array}
$$

Step-3:

$$
D^{(3)} = \begin{array}{c} \\ 1 \\ 2 \\ 3 \\ 4 \\ 5 \end{array}
\begin{array}{ccccc}
1 & 2 & 3 & 4 & 5 \\
\left[\begin{array}{ccccc}
0 & 3 & 10 & 8 & 25 \\
3 & 0 & 13 & 5 & 28 \\
10 & 13 & 0 & 6 & 15 \\
8 & 5 & 6 & 0 & 4 \\
\infty & \infty & \infty & 4 & 0
\end{array}\right]
\end{array},
\quad
S^{(3)} = \begin{array}{c} \\ 1 \\ 2 \\ 3 \\ 4 \\ 5 \end{array}
\begin{array}{ccccc}
1 & 2 & 3 & 4 & 5 \\
\left[\begin{array}{ccccc}
0 & 0 & 0 & 2 & 3 \\
0 & 0 & 1 & 0 & 3 \\
0 & 1 & 0 & 0 & 0 \\
2 & 0 & 0 & 0 & 0 \\
0 & 0 & 0 & 0 & 0
\end{array}\right]
\end{array}
$$

Step-4:

$$
D^{(4)} = \begin{array}{c} \\ 1 \\ 2 \\ 3 \\ 4 \\ 5 \end{array}
\begin{array}{ccccc}
1 & 2 & 3 & 4 & 5 \\
\left[\begin{array}{ccccc}
0 & 3 & 10 & 8 & 12 \\
3 & 0 & 11 & 5 & 9 \\
10 & 11 & 0 & 6 & 10 \\
8 & 5 & 6 & 0 & 4 \\
12 & 9 & 10 & 4 & 0
\end{array}\right]
\end{array},
\quad
S^{(4)} = \begin{array}{c} \\ 1 \\ 2 \\ 3 \\ 4 \\ 5 \end{array}
\begin{array}{ccccc}
1 & 2 & 3 & 4 & 5 \\
\left[\begin{array}{ccccc}
0 & 0 & 0 & 2 & 4 \\
0 & 0 & 4 & 0 & 4 \\
0 & 4 & 0 & 0 & 4 \\
2 & 0 & 0 & 0 & 0 \\
4 & 4 & 4 & 0 & 0
\end{array}\right]
\end{array}
$$

Step-5:

$$
D^{(5)} = \begin{array}{c} \\ 1 \\ 2 \\ 3 \\ 4 \\ 5 \end{array}
\begin{array}{ccccc}
1 & 2 & 3 & 4 & 5 \\
\left[\begin{array}{ccccc}
0 & 3 & 10 & 8 & 12 \\
3 & 0 & 11 & 5 & 9 \\
10 & 11 & 0 & 6 & 10 \\
8 & 5 & 6 & 0 & 4 \\
12 & 9 & 10 & 4 & 0
\end{array}\right]
\end{array},
\quad
S^{(5)} = \begin{array}{c} \\ 1 \\ 2 \\ 3 \\ 4 \\ 5 \end{array}
\begin{array}{ccccc}
1 & 2 & 3 & 4 & 5 \\
\left[\begin{array}{ccccc}
0 & 0 & 0 & 2 & 4 \\
0 & 0 & 4 & 0 & 4 \\
0 & 4 & 0 & 0 & 4 \\
2 & 0 & 0 & 0 & 0 \\
4 & 4 & 4 & 0 & 0
\end{array}\right]
\end{array}
$$

Now, from matrices $D^{(5)}$ and $S^{(5)}$, we can now determine the shortest route between every pair of nodes. For instance, let us consider the nodes 1 and 5.

The shortest distance from node 1 to node 5 is $d_{15}^{(5)} = 12$.

To determine the associated route, we determine the intermediate node $k = s_{15}^{(5)} = 4$, which yields the route 1–4–5.

Again, we determine the intermediate node $k = s_{14}^{(5)} = 2$ between nodes 1 and 4, yields the route 1–2–4. Since, $s_{12}^{(5)} = 0$, no further intermediate node exists between nodes 1 and 2. Similarly, no intermediate node exists between nodes 2 and 4.

Again, since, $s_{45}^{(5)} = 0$, no further intermediate node exists between nodes 4 and 5.

The combined result now gives the shortest route as 1–2–4–5. The associated length of the route is 12 units.

Example 5.5 Using *Floyd-Warshall's Algorithm*, determine the shortest route between all pair of vertices of the following Network (Fig 5.6).

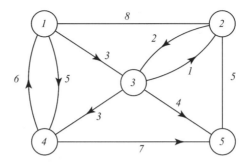

Fig. 5.6

Solution:

Initial step:

$$
D^{(0)} = \begin{array}{c} \\ 1 \\ 2 \\ 3 \\ 4 \\ 5 \end{array}
\begin{array}{c} \begin{array}{ccccc} 1 & 2 & 3 & 4 & 5 \end{array} \\
\left[\begin{array}{ccccc}
0 & 8 & 3 & 5 & \infty \\
8 & 0 & 2 & \infty & 5 \\
\infty & 1 & 0 & 3 & 4 \\
6 & \infty & \infty & 0 & 7 \\
\infty & 5 & \infty & \infty & 0
\end{array}\right] \end{array},
\quad
S^{(0)} = \begin{array}{c} \\ 1 \\ 2 \\ 3 \\ 4 \\ 5 \end{array}
\begin{array}{c} \begin{array}{ccccc} 1 & 2 & 3 & 4 & 5 \end{array} \\
\left[\begin{array}{ccccc}
0 & 0 & 0 & 0 & 0 \\
0 & 0 & 0 & 0 & 0 \\
0 & 0 & 0 & 0 & 0 \\
0 & 0 & 0 & 0 & 0 \\
0 & 0 & 0 & 0 & 0
\end{array}\right] \end{array}
$$

At each Step $k(k = 1, 2 \ldots, n = 5)$, node k is considered as intermediate node between every pair of nodes. The kth row, kth column, and diagonal elements of $D^{(k-1)}$ and $S^{(k-1)}$ will remain unchanged in Step k.

Step-1:

$$
D^{(1)} = \begin{array}{c} \\ 1 \\ 2 \\ 3 \\ 4 \\ 5 \end{array}
\begin{array}{c} \begin{array}{ccccc} 1 & 2 & 3 & 4 & 5 \end{array} \\
\left[\begin{array}{ccccc}
0 & 8 & 3 & 5 & \infty \\
8 & 0 & 2 & 13 & 5 \\
\infty & 1 & 0 & 3 & 4 \\
6 & 14 & 9 & 0 & 7 \\
\infty & 5 & \infty & \infty & 0
\end{array}\right] \end{array},
\quad
S^{(1)} = \begin{array}{c} \\ 1 \\ 2 \\ 3 \\ 4 \\ 5 \end{array}
\begin{array}{c} \begin{array}{ccccc} 1 & 2 & 3 & 4 & 5 \end{array} \\
\left[\begin{array}{ccccc}
0 & 0 & 0 & 0 & 0 \\
0 & 0 & 0 & 1 & 0 \\
0 & 0 & 0 & 0 & 0 \\
0 & 1 & 1 & 0 & 0 \\
0 & 0 & 0 & 0 & 0
\end{array}\right] \end{array}
$$

Step-2:

$$
D^{(2)} = \begin{array}{c} \\ 1 \\ 2 \\ 3 \\ 4 \\ 5 \end{array}
\begin{array}{c} \begin{array}{ccccc} 1 & 2 & 3 & 4 & 5 \end{array} \\
\left[\begin{array}{ccccc}
0 & 8 & 3 & 5 & 13 \\
8 & 0 & 2 & 13 & 5 \\
9 & 1 & 0 & 3 & 4 \\
6 & 14 & 9 & 0 & 7 \\
13 & 5 & 7 & 18 & 0
\end{array}\right] \end{array},
\quad
S^{(2)} = \begin{array}{c} \\ 1 \\ 2 \\ 3 \\ 4 \\ 5 \end{array}
\begin{array}{c} \begin{array}{ccccc} 1 & 2 & 3 & 4 & 5 \end{array} \\
\left[\begin{array}{ccccc}
0 & 0 & 0 & 0 & 2 \\
0 & 0 & 0 & 1 & 0 \\
2 & 0 & 0 & 0 & 0 \\
0 & 1 & 1 & 0 & 0 \\
2 & 0 & 2 & 2 & 0
\end{array}\right] \end{array}
$$

Step-3:

$$
D^{(3)} = \begin{array}{c} \\ 1 \\ 2 \\ 3 \\ 4 \\ 5 \end{array}
\begin{array}{ccccc} 1 & 2 & 3 & 4 & 5 \\ \left[\begin{array}{ccccc} 0 & 4 & 3 & 5 & 7 \\ 8 & 0 & 2 & 5 & 5 \\ 9 & 1 & 0 & 3 & 4 \\ 6 & 10 & 9 & 0 & 7 \\ 13 & 5 & 7 & 10 & 0 \end{array}\right] \end{array},
\quad
S^{(3)} = \begin{array}{c} \\ 1 \\ 2 \\ 3 \\ 4 \\ 5 \end{array}
\begin{array}{ccccc} 1 & 2 & 3 & 4 & 5 \\ \left[\begin{array}{ccccc} 0 & 3 & 0 & 0 & 3 \\ 0 & 0 & 0 & 3 & 0 \\ 2 & 0 & 0 & 0 & 0 \\ 0 & 3 & 1 & 0 & 0 \\ 2 & 0 & 2 & 3 & 0 \end{array}\right] \end{array}
$$

Step-4:

$$
D^{(4)} = \begin{array}{c} \\ 1 \\ 2 \\ 3 \\ 4 \\ 5 \end{array}
\begin{array}{ccccc} 1 & 2 & 3 & 4 & 5 \\ \left[\begin{array}{ccccc} 0 & 4 & 3 & 5 & 7 \\ 8 & 0 & 2 & 5 & 5 \\ 9 & 1 & 0 & 3 & 4 \\ 6 & 10 & 9 & 0 & 7 \\ 13 & 5 & 7 & 10 & 0 \end{array}\right] \end{array},
\quad
S^{(4)} = \begin{array}{c} \\ 1 \\ 2 \\ 3 \\ 4 \\ 5 \end{array}
\begin{array}{ccccc} 1 & 2 & 3 & 4 & 5 \\ \left[\begin{array}{ccccc} 0 & 3 & 0 & 0 & 3 \\ 0 & 0 & 0 & 3 & 0 \\ 2 & 0 & 0 & 0 & 0 \\ 0 & 3 & 1 & 0 & 0 \\ 2 & 0 & 2 & 3 & 0 \end{array}\right] \end{array}
$$

Step-5:

$$
D^{(5)} = \begin{array}{c} \\ 1 \\ 2 \\ 3 \\ 4 \\ 5 \end{array}
\begin{array}{ccccc} 1 & 2 & 3 & 4 & 5 \\ \left[\begin{array}{ccccc} 0 & 4 & 3 & 5 & 7 \\ 8 & 0 & 2 & 5 & 5 \\ 9 & 1 & 0 & 3 & 4 \\ 6 & 10 & 9 & 0 & 7 \\ 13 & 5 & 7 & 10 & 0 \end{array}\right] \end{array},
\quad
S^{(5)} = \begin{array}{c} \\ 1 \\ 2 \\ 3 \\ 4 \\ 5 \end{array}
\begin{array}{ccccc} 1 & 2 & 3 & 4 & 5 \\ \left[\begin{array}{ccccc} 0 & 3 & 0 & 0 & 3 \\ 0 & 0 & 0 & 3 & 0 \\ 2 & 0 & 0 & 0 & 0 \\ 0 & 3 & 1 & 0 & 0 \\ 2 & 0 & 2 & 3 & 0 \end{array}\right] \end{array}
$$

The shortest route between all pair of vertices of the given network can be obtained from the above matrices $D^{(5)}$ and $S^{(5)}$, where the shortest distance between each pair of vertices can be found from $D^{(5)}$ and the shortest path can be found from $S^{(5)}$ respectively. Table 5.22 shows the shortest route between all pair of vertices.

Table 5.22

Starting node to destination node	Shortest distance	Shortest path
1–2	4	1–3–2
1–3	3	1–3
1–4	5	1–4
1–5	7	1–3–5
2–1	8	2–1
2–3	2	2–3
2–4	5	2–3–4
2–5	5	2–5

(continued)

Table 5.22 (continued)

Starting node to destination node	Shortest distance	Shortest path
3–1	9	3–2–1
3–2	1	3–2
3–4	3	3–4
3–5	4	3–5
4–1	6	4–1
4–2	10	4–1–3–2
4–3	9	4–1–3
4–5	7	4–5
5–1	13	5–2–1
5–2	5	5–2
5–3	7	5–2–3
5–4	10	5–2–3–4

5.1.2.3 Comparison Between Floyd-Warshall's Algorithm with Dijkstra's Algorithm

The all-pairs-shortest-path problem is generalization of the single-source-shortest path problem, so we can use Floyd's Algorithm or Dijkstra's Algorithm (Varying the source node over all nodes)

1. The time complexity of Floyd's Algorithm is $O\left(|V|^3\right)$ i.e. $O(n^3)$, where $|V| = n$.
2. The time complexity of Dijkstra's Algorithm with an adjacency matrix is $O(n^2)$. So, varying over n source nodes, it is $O(n^3)$.
3. The time complexity of Dijkstra's Algorithm with an adjacency list (the representation of all edges in a graph as a list) is $O(E.\log|V|)$. So, varying over n source nodes, it is $O(|V|E\log|V|)$.

For large sparse graph, Dijkstra's Algorithm is preferable.

5.2 Minimum Spanning Tree Problem

Consider the following example of laying telephone cable in a locality as shown in Fig. 5.7a.

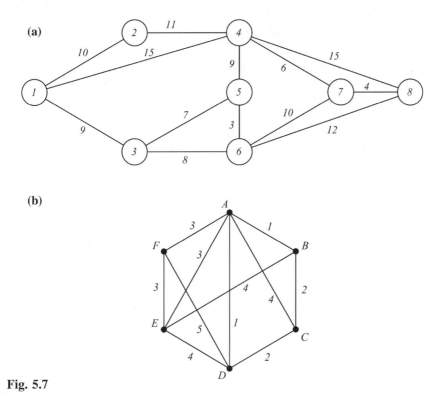

Fig. 5.7

Figure 5.7a summarizes the distance network of the locality. The number on each edge represents the distance between the nodes connected by that edge.

5.2.1 Objective of Minimum Spanning Tree Problem

The objective of the minimum spanning tree problem is to connect the nodes of the network by a set of edges such that the total length of the edges is minimized. In process of constructing the minimum spanning tree, it should be taken care that there is no cycle in it. With reference to the telephone cable laying example, the objective is to connect all the nodes by a set of edges such that the total length of the telephone cable to be laid is minimized.

In this section, the following algorithms for the minimum spanning tree problem are presented,

- Prim's Algorithm
- Kruskal's Algorithm

5.2.2 *Minimum Spanning Tree*

Let G be a weighted graph in which each edge e has been assigned a real number $w(e)$, called the weight of the edge e. If H be a subgraph of a weighted graph, the weight $w(H)$ of H, is the sum of the weights $w(e_1) + w(e_2) + \cdots + w(e_k)$ where $\{e_1, e_2, \ldots, e_k\}$ is the set of edges of H.

A spanning tree T of a weighted graph G is called a minimal spanning tree if its weight is minimum. That is $w(T)$ is minimum where $w(T) = w(e_1) + w(e_2) + \cdots + w(e_k)$ and $\{e_1, e_2, \ldots, e_k\}$ is the set of edges of T.

Many optimization problems involves finding in a suitable weighted graph, a certain type of subgraph with minimum weight. To illustrate, let G be the graph whose vertex set is the set of cities and in which uv be an edge if and only if it is possible to build a pipeline joining the cities u and v. We can then consider G as a weighted graph by assigning to each edge the cost of constructing the corresponding pipeline. For example, suppose that there are six cities A, B, C, D, E, F and we get the weighted graph G as shown in Fig. 5.7b.

Absence of an edge from B to D indicates that it is not possible to build a pipeline from B to D. The number (weight) five assigned to the edge from F to D indicates the cost of building a pipeline from F to D.

Since the problem is to ensure that every city is supplied with water from the source city, we are looking for a connected spanning subgraph of G. Moreover, since we want to do this in the most economical way, such a spanning subgraph should have no cyles, because the deletion of an edge (a pipeline) from a cycle in a connected spanning subgraph still leaves us with a connected spanning subgraph. Therefore, we are looking for a spanning tree of G. Moreover the economical factor implies that we want the cheapest such spanning tree, i.e., a spanning tree with minimum weight.

Here, we now present two algorithms, due to Kruskal and Prim, for finding a minimal spanning tree for a connected weighted graph where no weight is negative.

5.2.2.1 Kruskal's Algorithm

Let $G = (V, E)$ be a weighted connected graph.

Step-1: Select one edge e_i of G such that its weight $w(e_i)$ is minimum.

Step-2:

1. If the edges $e_1, e_2, \ldots, e_k$ have been chosen then select an edge e_{k+1} such that $e_{k+1} \neq e_i$ for $i = 1, 2, \ldots, k$
2. The edges $e_1, e_2, \ldots, e_k, e_{k+1}$ does not form a circuit.
3. The weight of $w(e_{k+1})$ is as small as possible subject to the condition number 2 of step-2 above.

Step-3:

Stop, when all the vertices of G are in T which is the required spanning tree of G with $n - 1$ edges.

Example 5.6 Use Kruskal's Algorithm, to find the minimum spanning tree of the graph in Fig. 5.8

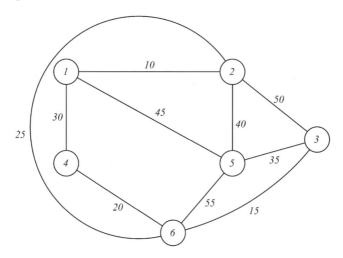

Fig. 5.8

Solution:

Sl. number	Edge	Weight	Corresponding graph
1	(1,2)	10	
2	(6,3)	15	
3	(4,6)	20	
4	(2,6)	25	

(continued)

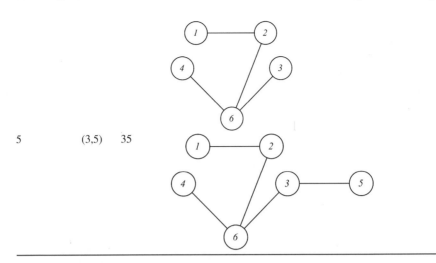

5   (3,5)   35

The required Minimal Spanning tree is

Fig. 5.9 A minimal
spanning tree

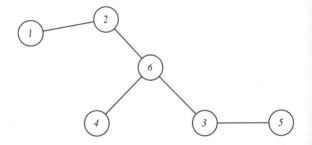

Total Weight $= 10 + 15 + 20 + 25 + 35 = 105$ units which is minimum weight of the Spanning Tree in Fig. 5.9.

Example 5.7 Apply Kruskal's Algorithm, to find the minimum spanning tree of the graph in Fig. 5.10

Fig. 5.10

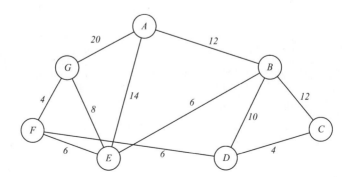

Solution:

Sl. number	Edge	Weight	Corresponding graph
1	FG	4	
2	CD	4	
3	FD	6	
4	EB	6	
5	EF	6	
6	AB	12	

Total Weight $= 4 + 4 + 6 + 6 + 6 + 12 = 38$ units which is minimum weight of the Spanning Tree appears in Sl. No. 6.

5.2.2.2 Prim's Algorithm

Let T be a tree in a connected weighted graph G represented by two sets: the set of vertices in T and set of edges in T.

Step-1: Start with a vertex v_0 (say) in G and no edge such that $T = \{\{v_0\}, \phi\}$

Step-2: Find the edge $e_1 = (v_0, v_1)$ in G such that the end vertex v_0 is in T and its weight is minimum, i.e., $w(e_1)$ is minimum. Adjoin v_1 and e_1 to T, i.e., $T = \{\{v_0, v_1\}, e_1\}$.

Step-3: Choose the next edge $e_{ij} = (v_i, v_j)$ in such a way that end vertex v_i is in T and end vertex v_j is not in T and weight of e_{ij} is as small as possible. Adjoin v_j and e_{ij} to T.

Step-4: Repeat step-3 until T contains all the vertices of G. The set T will give minimal spanning tree of G.

Example 5.8 Apply Prim's Algorithm, to find the minimum spanning tree of the graph in Fig. 5.10.

Solution:

The Adjacency matrix $X(G)$, see Chap. 6, is

$$X(G) = \begin{array}{c} \\ A \\ B \\ C \\ D \\ E \\ F \\ G \end{array} \begin{array}{c} \begin{array}{ccccccc} A & B & C & D & E & F & G \end{array} \\ \left[\begin{array}{ccccccc} 0 & 12 & \infty & \infty & 14 & \infty & 20 \\ 12 & 0 & 12 & 10 & 6 & \infty & \infty \\ \infty & 12 & 0 & 4 & \infty & \infty & \infty \\ \infty & 10 & 4 & 0 & \infty & 6 & \infty \\ 14 & 6 & \infty & \infty & 0 & 6 & 8 \\ \infty & \infty & \infty & 6 & 6 & 0 & 4 \\ 20 & \infty & \infty & \infty & 8 & 4 & 0 \end{array} \right] \end{array}$$

Sl. no.	Tree T	Corresponding tree T
1	$T = \{\{A\}, \phi\}$ Minimum weight in the row of A is 12 *corresponding to the column of vertex B*. We include vertex B and edge (A, B) to T	A •
2	$T = \{\{A, B\}, \{(A, B)\}\}$ Minimum weight in the rows of A and B is 6 *corresponding to the column of vertex E*. We include vertex E and edge (B, E) to T	A •———————• B

(continued)

(continued)

Sl. no.	Tree T	Corresponding tree T
3	$T = \{\{A,B,E\}, \{(A,B),(B,E)\}\}$	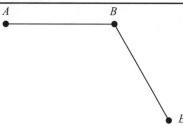

Minimum weight in the rows of A, B and E is 6 occurs in the row of *corresponding to the column of vertex F*. We include vertex F and edge (E,F) to T

4	$T = \{\{A,B,E,F\}, \{(A,B),(B,E),(E,F)\}\}$	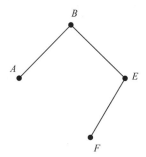

Minimum weight in the rows of A, B, E and F is 4 occurs in the row of F *corresponding to the column of vertex G*. We include vertex G and edge (F,G) to T

5	$T = \{\{A,B,E,F,G\}, \{(A,B), (B,E),(E,F),(F,G)\}\}$	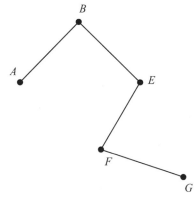

Minimum weight in the rows of A, B, E, F and G is 4. But the corresponding vertices F and G are already present in T. The next minimum is 6 occurs in the row of F *corresponding to the column of vertex D*. We include vertex D and edge (F,D) to T

(continued)

(continued)

Sl. no.	Tree T	Corresponding tree T
6	$T = \{\{A, B, D, E, F, G\}, \{(A, B), (B, E), (E, F), (F, G), (F, D)\}\}$	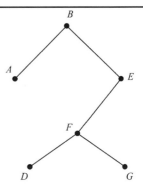
	Minimum weight in the rows of $A, B, D, E, F,$ and G is 4, in the row of D corresponding to the column of vertex C. We include vertex C and edge (D, C) to T.	
7	$T = \{\{A, B, C, D, E, F, G\}, \{(A, B), (B, E), (E, F), (F, G), (F, D), (D, C)\}\}$	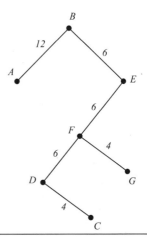

Total Weight $= 12 + 6 + 6 + 4 + 6 + 4 = 38$ units which is the minimum weight of the Spanning Tree appears in Sl. No. 7.

Example 5.9 Apply Prim's Algorithm, to find the minimum spanning tree of the graph in Fig. 5.11.

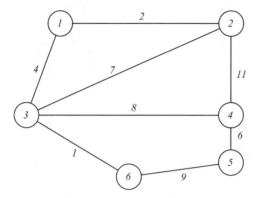

Fig. 5.11

Solution:

The Adjacency matrix $X(G)$ is

$$X(G) = \begin{array}{c} \\ 1 \\ 2 \\ 3 \\ 4 \\ 5 \\ 6 \end{array} \begin{array}{c} \begin{array}{cccccc} 1 & 2 & 3 & 4 & 5 & 6 \end{array} \\ \left[\begin{array}{cccccc} 0 & 2 & 4 & \infty & \infty & \infty \\ 2 & 0 & 7 & 11 & \infty & \infty \\ 4 & 7 & 0 & 8 & \infty & 1 \\ \infty & 11 & 8 & 0 & 6 & \infty \\ \infty & \infty & \infty & 6 & 0 & 9 \\ \infty & \infty & 1 & \infty & 9 & 0 \end{array} \right] \end{array}$$

Sl. no.	Tree T	Corresponding tree T
1	$T = \{\{1\}, \phi\}$	 ①
	Minimum weight in the row of 1 is 2 *corresponding to the column of vertex* 2. We include vertex 2 and edge (1, 2) to T.	
2	$T = \{\{1, 2\}, \{(1, 2)\}\}$	
	Minimum weight in the rows of 1 and 2 is 4 (next to 2) in the row of 1, *corresponding to the column of vertex* 3. We include vertex 3 and edge (1, 3) to T.	

(continued)

(continued)

Sl. no.	Tree T	Corresponding tree T
3	$T = \{\{1,2,3\},\{(1,2),(1,3)\}\}$	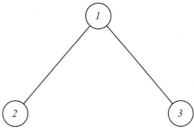

Minimum weight in the rows of
1, 2, and 3 is 1 in the row of 3,
corresponding to the column of vertex 6.
We include vertex 6 and edge (3, 6) to T.

| 4 | $T = \{\{1,2,3,6\},\{(1,2),(1,3),(3,6)\}\}$ | 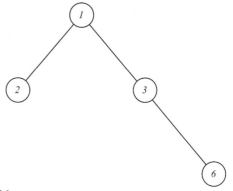 |

Minimum weight in the rows of 1, 2,3, and 6
is 8 in the row of 3, *corresponding
to the column of vertex* 4. Since, the
other vertices corresponding to the
minimum than 8 are already
present in T.
We include vertex 4 and
edge (3,4) to T.

(continued)

(continued)

Sl. no.	Tree T	Corresponding tree T
5	$T = \{\{1,2,3,6,4\}, \{(1,2), (1,3), (3,6), (3,4)\}\}$	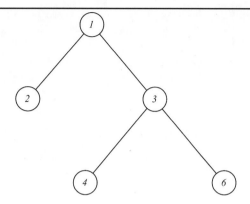
	Minimum weight in the rows of 1, 2, 3, 6, and 4 is 6 in the row of 4, *corresponding to the column of vertex* 5. Since, the other vertices corresponding to the minimum than 6 are already present in T. We include vertex 5 and edge (4, 5) to T.	
6	$T = \{\{1,2,3,6,4,5\}, \{(1,2),(1,3),(3,6), (3,4),(4,5)\}\}$	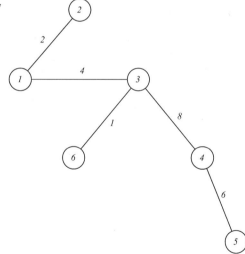

Total Weight $= 2 + 4 + 1 + 8 + 6 = 21$ units which is minimum weight of the Spanning Tree appears in Sl. No. 6.

5.3 Breadth First Search Algorithm to Find the Shortest Path

Let G be an unweighted graph. We can find the shortest path from v_k to v_p.

Input: Connected graph $G = (V, E)$ in which one vertex is denoted by v_k and one by v_p and each edge (i, j) has length $l_{ij} = 1$. Initially, all vertices are unlabeled.

Output: A shortest path $v_k \rightarrow v_p$ in $G = (V, E)$.

Step 1: Label the starting vertex v_k with 0.

Step 2: Set $i = 0$.

Step 3: Find all unlabeled vertices in G which are adjacent to the vertices labeled i. If there are no such vertices then $\exists$ no path from v_k to v_p. Otherwise go to step 4.

Step 4: Label the vertices just found with $i + 1$.

Step 5: If vertex v_p is labeled. Stop. The value of the label of v_p is the shortest distance from v_k to v_p. Otherwise go to step 3.

Now, we use the following backtracking process to find the shortest path from v_p to v_k. Let the destination vertex v_p is labeled $r + 1$. Then find a vertex adjacent to v_p whose label is r. Continue this process until the initial vertex v_k is reached. $\square$

Example 5.10 Find by BFS method the shortest path from the vertex v_2 to v_6 in the following graph (Fig 5.12).

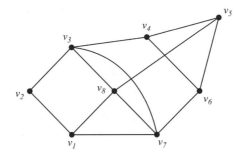

Fig. 5.12

Solution:

According to BFS Algorithm, the shortest path is $v_2 - v_1 - v_7 - v_6$. So, the shortest distance from v_2 to v_6 is 3.

Figure 5.13 shows the Breadth first tree which is also a spanning tree. $v_2v_1v_3v_7v_8v_4v_6v_5$ is called Breadth first traversal.

Fig. 5.13 Breadth first tree

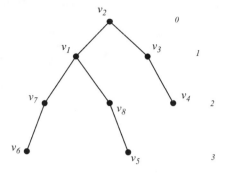

5.3.1 BFS Algorithm for Construction of a Spanning Tree

Discard all the parallels and loops from the given graph.

Choose any vertex v_k of the graph. In this algorithm, label this vertex as 0. Then we proceed stage by stage by labeling a new vertex at every stage according to the following rule:

Find all unlabeled vertex in G which is adjacent to the vertices labeled i. Label those vertices as $i + 1$ and get them joined (with i labeled vertex) by edges so that no circuit is formed. This stage to stage labeling and joining stops when all vertices are labeled. All the vertices and the successive joining edges form the required spanning tree.   □

Example 5.11 Find by BFS algorithm a spanning tree in the following graph (Fig 5.14).

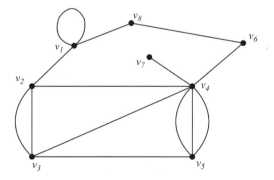

Fig. 5.14

Solution:
After discarding the self-loop and the parallel edges, the resultant graph is shown in Fig. 5.15.

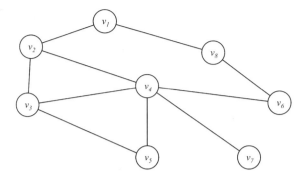

Fig. 5.15

Using BFS Algorithm, a breadth First Tree (Spanning Tree) has been obtained as shown in Fig. 5.16.

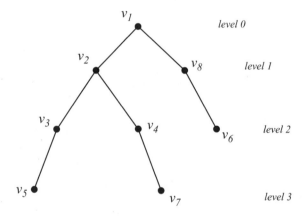

Breadth First Tree (Spanning tree)

Fig. 5.16

5.4 Depth First Search Algorithm for Construction of a Spanning Tree

Discard all parallels and loops from the given graph.

Choose any vertex v_k of the resultant graph. Make a path starting from v_k as long as possible by successively adding edges and vertices. Let this path be $P_1 : v_k \rightarrow v_p$.

Now backtrack (along P_1) from v_p and Let v_a be the first vertex starting from which we can make a path (as long as possible) as P_2 containing no vertices of P_1 so that it does not form a circuit.

Let v_b be the next vertex reached along this tracking from which we can make another path (as long as possible) say P_3 containing no vertices of P_1 and P_2. Continuing this process, we get paths $P_4, P_5, \ldots$ and so on. We stop at that stage when each of the vertices of the graph is included in some of the paths $P_1, P_2, P_3, \ldots$. These paths together form the required spanning tree. $\square$

Example 5.12 Find by DFS algorithm a spanning tree in the following graph (Fig 5.17).

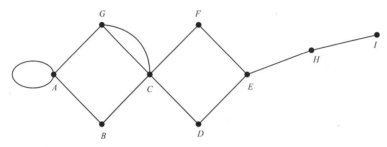

Fig. 5.17

Solution:
After discarding all loops and parallels, we have the following graph in Fig. 5.18.

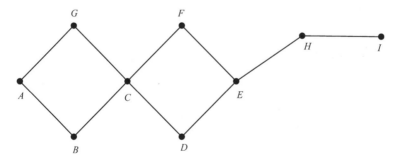

Fig. 5.18

Choose arbitrarily the vertex G. Make a path starting from G as long as possible by successively adding edges and vertices. This path be $P_1 : G - C - F - E - H - I$. Now backtrack from I to H. We get no path starting from H. Next we backtrack from H to E, we get a path $P_2 : E - D$, noting that P_2 does not make any circuit with the path P_1 (Fig. 5.19).

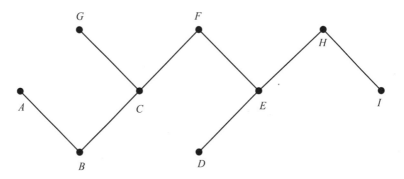

Fig. 5.19 Spanning tree obtained by DFS

Again, from E we re-track to F and then to C. There exists a path from C, say, $P_3 : C - B - A$ so that all of its vertices are not included in the previous two paths. Since the three paths P_1, P_2 and P_3 contain all the vertices of the graph, we can stop at this stage.

Example 5.13 Find by DFS the spanning tree of the following graph (Fig. 5.20).

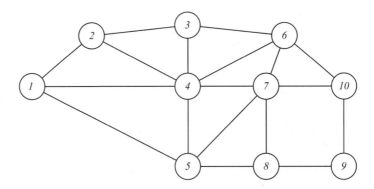

Fig. 5.20

Solution:
In Fig. 5.21, all the vertices are included in the path P: $1 - 2 - 3 - 6 - 10 - 9 - 8 - 7 - 5 - 4$ which is the required spanning tree obtained by DFS.

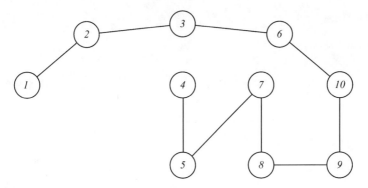

Fig. 5.21 Spanning tree obtained by DFS

Example 5.14 Find the fundamental circuits of the following graph

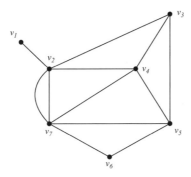

Fig. 5.22

Solution:
After discarding the parallel edge of Fig. 5.22, we obtain the following graph in Fig. 5.23.

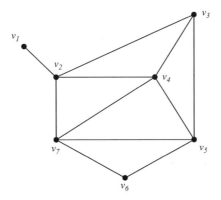

Fig. 5.23

Fig. 5.24 Fundamental circuits represented by five dotted chord lines

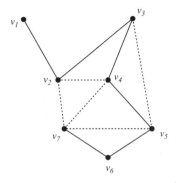

Figure 5.24 shows the spanning tree obtained by DFS.

From the simple graph Fig. 5.23, we see that there are $e - n + 1 = 11 - 7 + 1 = 5$ fundamental circuits with regard to the chords $v_2 - v_4$, $v_3 - v_5$, $v_2 - v_7$, $v_4 - v_7$ and $v_7 - v_5$ which are shown in Fig. 5.24 by dotted lines.

	Fundamental circuit	Corresponding chord
1	$v_2 - v_3 - v_4 - v_2$	$v_2 - v_4$
2	$v_3 - v_4 - v_5 - v_3$	$v_3 - v_5$
3	$v_2 - v_3 - v_4 - v_5 - v_6 - v_7 - v_2$	$v_2 - v_7$
4	$v_4 - v_5 - v_6 - v_7 - v_4$	$v_4 - v_7$
5	$v_7 - v_6 - v_5 - v_7$	$v_7 - v_5$

Example 5.15 Find the fundamental circuits of the following graph.

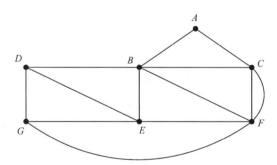

Fig. 5.25

Solution:

After discarding the parallel edge of Fig. 5.25, we obtain the following graph in Fig. 5.26.

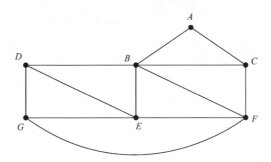

Fig. 5.26

Figure 5.27 shows the Spanning tree obtained by DFS viz. the path P:
$A - B - D - G - E - F - C$.

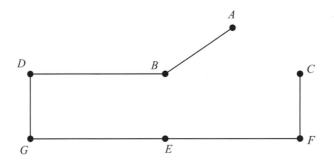

Fig. 5.27 Spanning tree obtained by DFS

From the simple graph Fig. 5.26, we see that there are $e - n + 1 = 12 - 7 + 1 = 6$ fundamental circuits with regard to the chords AC, BC, BF, BE, DE, and GF.

Note: Let T be a spanning tree in a connected graph G. *Adding any one chord to T will create exactly one circuit*. Such a circuit, formed by adding a chord to a spanning tree, is called a *fundamental circuit*.

Exercises:

1. Apply Dijkstra's Algorithm, to find the shortest route from node 3 to 4 of the following network in Fig. 5.28.

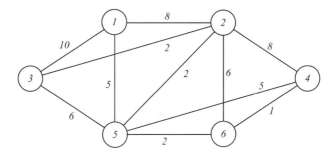

Fig. 5.28

2. Use Dijkstra's Algorithm, to determine the shortest route between the following cities: (Fig. 5.29)

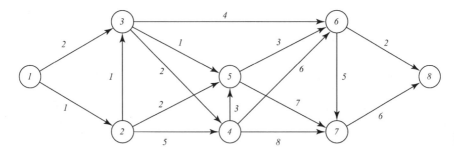

Fig. 5.29

(a) Cities 1 and 8.
(b) Cities 4 and 8.
(c) Cities 2 and 6.

3. Apply Dijkstra's Algorithm, to find the shortest route between node 1 and each of the remaining nodes (Fig. 5.30).

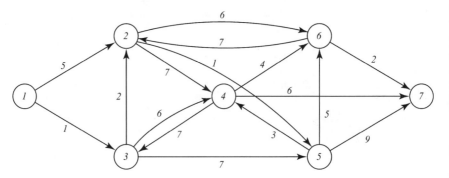

Fig. 5.30

4. Apply *Breadth First Search* (BFS) Algorithm, to find a Spanning tree of the following graph and hence find all *Fundamental Circuits* for the following graph in Fig. 5.31.

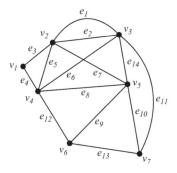

Fig. 5.31

5. Apply *Depth First Search* (DFS) Algorithm, to find a Spanning tree of the following graph and hence find all *Fundamental Cut sets* for the graph in Fig. 5.31.
6. By BFS Algorithm, find a shortest path from vertex A to Z in the following two graphs (Fig. 5.32).

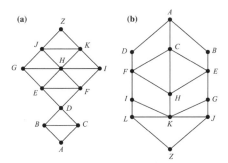

Fig. 5.32

7. Find by Kruskal's Algorithm a minimal spanning tree from the following graph G (Fig. 5.33).

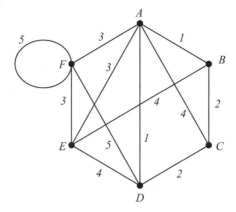

Fig. 5.33

8. Use DFS Algorithm to find a spanning tree of the following graph (Fig. 5.34).

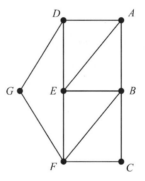

Fig. 5.34

9. Table 5.23 shows the distances, in kilometers, between six villages in India.
   Find a minimal spanning tree connecting the six villages using Prim's
   Algorithm.

Table 5.23

	A	B	C	D	E	F
A	–	5	6	12	4	7
B	5	–	11	3	2	5
C	6	11	–	8	6	6
D	12	3	8	–	7	9
E	4	2	6	7	–	8
F	7	5	6	9	8	–

10. By Prim's Algorithm, find a minimal spanning tree in the following graphs and find the corresponding minimum weight. (Fig. 5.35).

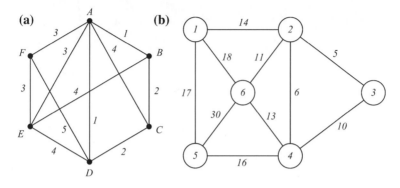

Fig. 5.35

11. Find the shortest distance matrix and the corresponding shortest path matrix for all the pairs of vertices in the directed weighted graph given in Fig. 5.36, using Floyd-Warshall's algorithm.

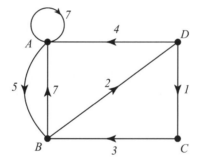

Fig. 5.36

12. Use Prim's Algorithm to find a minimum spanning tree for the weighted graph
    given in Fig. 5.37.

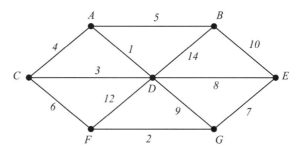

Fig. 5.37

13. Find the minimum spanning tree for the weighted graph shown in Fig. 5.38,
    by using Kruskal's algorithm.

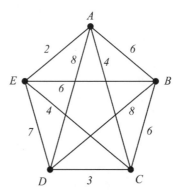

Fig. 5.38

14. Use Kruskal's algorithm, to find a minimum spanning tree for the weighted
    graph as shown in Fig. 5.39.

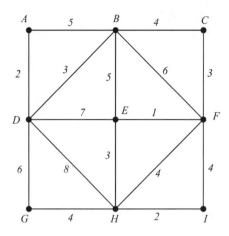

Fig. 5.39

15. Consider the network given below, to find the minimum spanning tree using Prim's Algorithm (Fig. 5.40).

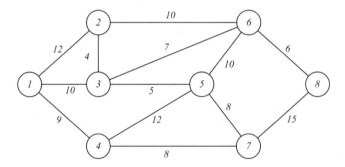

Fig. 5.40

16. Consider the following network and find the minimum spanning tree using
    (a) The Prim's Algorithm
    (b) The Kruskal's Algorithm (Fig. 5.41)

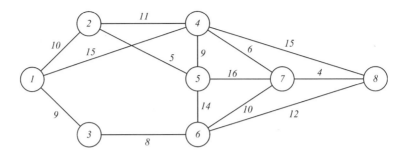

Fig. 5.41

17. Find the minimum spanning tree of the following network using Kruskal's Algorithm (Fig. 5.42).

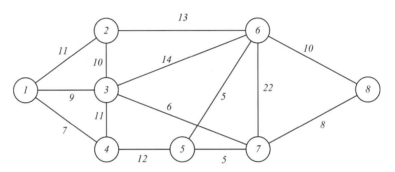

Fig. 5.42

18. Prove that if G is a connected weighted graph in which no two edges have the same weight then G has a unique minimum spanning tree.
19. Using Floyd-Warshall's Algorithm determine the shortest route between all pair of vertices of the following graph (Fig. 5.43).

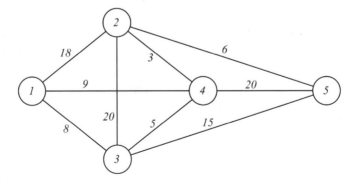

Fig. 5.43

20. Using Floyd-Warshall's Algorithm determine the shortest route between all pair of vertices of the following graph (Fig. 5.44).

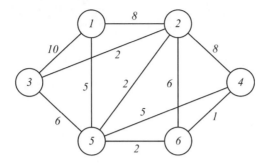

Fig. 5.44

21. By using Kruskal's Algorithm, find a Minimal Spanning Tree in the following graph and find the corresponding minimum weight (Fig. 5.45).

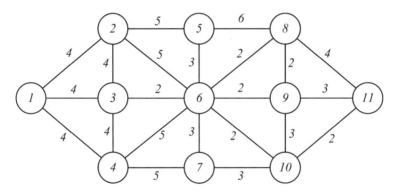

Fig. 5.45

Chapter 6
Matrix Representation on Graphs

6.1 Vector Space Associated with a Graph

Let us consider a graph G in Fig. 6.1 with four vertices and five edges e_1, e_2, e_3, e_4, e_5. Any subgraph H of G can be represented by a 5-tuple.

$$X = (x_1, x_2, x_3, x_4, x_5)$$

such that

$$x_i = 1, \text{ if } e_i \text{ is in } H$$
$$\text{and } x_i = 0, \text{ if } e_i \text{ is not in } H$$

For instance, the subgraph H_1 in Fig. 6.1 will be represented by $(1, 0, 1, 0, 1)$.

There are $2^5 = 32$ such 5-tuples possible, including the zero vector $\mathbf{0} = (0, 0, 0, 0, 0)$ which represents a null graph and $(1, 1, 1, 1, 1)$ which is G itself.

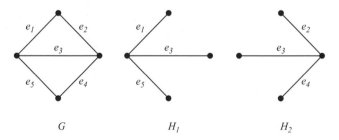

$$G \qquad\qquad H_1 \qquad\qquad H_2$$

Fig. 6.1 Graph G and its two subgraphs H_1 and H_2

The ring-sum operation between two subgraphs corresponds to the *modulo* 2 addition between the two 5-tuples representing the two subgraphs.

For example, consider two subgraphs

$$H_1 = \{e_1, e_3, e_5\} \text{ represented by } (1, 0, 1, 0, 1)$$
$$\text{and } H_2 = \{e_2, e_3, e_4\} \text{ represented by } (0, 1, 1, 1, 0)$$

S. Saha Ray, *Graph Theory with Algorithms and its Applications*,
DOI: 10.1007/978-81-322-0750-4_6, © Springer India 2013

The ring sum

$$H_1 \oplus H_2 = \{e_1, e_2, e_4, e_5\} \text{ represented by } (1, 1, 0, 1, 1)$$

which is clearly *modulo* 2 addition of the 5-tuples for H_1 and H_2.

There is a vector space W_G associated with every graph G and this vector space consists of

1. Galois field *modulo* 2 $(GF(2))$, i.e., the set $\{0, 1\}$ with operation addition *modulo* 2 and multiplication *modulo* 2.
2. 2^e vectors (e-tuples), where e is the number of edges of G.
3. An addition operation between two vectors $X = (x_1, x_2, \ldots, x_e), Y = (y_1, y_2, \ldots, y_e)$ in this space, defined by the vector sum

$$X \oplus Y = (x_1 + y_1, x_2 + y_2, \ldots, x_e + y_e)$$

where $+$ is the addition *modulo* 2.

4. And a scalar multiplication between a scalar c in $GF(2)$ and a vector $X = (x_1, x_2, \ldots, x_e)$ in this space, defined as

$$c.X = (c.x_1, c.x_2, \ldots, c.x_e).$$

where $\cdot$ is the multiplication *modulo* 2.

Basis vectors of a graph:

Let W_G be the vector space associated with a graph G. Corresponding to each subgraph of G, there exists a vector in W_G, represented by an e-tuple. The *standard basis* for this vector space W_G is a set of e linearly independent vectors, each representing a subgraph consisting of one edge of G. For instance, for the graph in Fig. 6.1, the set of the following five vectors forms a basis for W_G.

$$\{(1, 0, 0, 0, 0), (0, 1, 0, 0, 0), (0, 0, 1, 0, 0), (0, 0, 0, 1, 0), (0, 0, 0, 0, 1)\}$$

Any of the possible 32 subgraphs, including G itself as well as the null graph, can be represented by a suitable linear combination of these five basis vectors.

6.2 Matrix Representation of Graphs

6.2.1 Incidence Matrix

Let G be a graph with n vertices, e edges, and no self-loops. We define a matrix $A = (a_{ij})_{n \times e}$ where n rows correspond to the n vertices and the e columns correspond to the e edges, as follows:

$$a_{ij} = 1, \text{ if } j\text{th edge } e_j \text{ is incident on } i\text{th vertex } v_i$$
$$= 0, \text{ otherwise}$$

This matrix A is called **incidence matrix** of G. Sometimes it is written as $A(G)$.

Example 6.1 Find the incidence matrix of the following graph

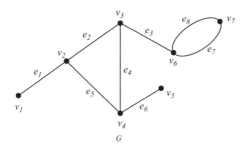

G

Fig. 6.2

Solution:
The incidence matrix $A(G)$ of the graph G in Fig. 6.2 is as follows:

$$A(G) = \begin{array}{c} \\ v_1 \\ v_2 \\ v_3 \\ v_4 \\ v_5 \\ v_6 \\ v_7 \end{array}
\begin{array}{c} e_1\ e_2\ e_3\ e_4\ e_5\ e_6\ e_7\ e_8 \\
\begin{bmatrix}
1 & 0 & 0 & 0 & 0 & 0 & 0 & 0 \\
1 & 1 & 0 & 0 & 1 & 0 & 0 & 0 \\
0 & 1 & 1 & 1 & 0 & 0 & 0 & 0 \\
0 & 0 & 0 & 1 & 1 & 1 & 0 & 0 \\
0 & 0 & 0 & 0 & 0 & 1 & 0 & 0 \\
0 & 0 & 1 & 0 & 0 & 0 & 1 & 1 \\
0 & 0 & 0 & 0 & 0 & 0 & 1 & 1
\end{bmatrix}
\end{array}$$

Example 6.2 Determine the graph where incidence matrix is

$$\begin{array}{c} \\ v_1 \\ v_2 \\ v_3 \\ v_4 \\ v_5 \end{array}
\begin{array}{c} e_1\ \ e_2\ e_3\ e_4\ e_5\ e_6 \\
\begin{bmatrix}
0 & 1 & 0 & 0 & 1 & 1 \\
1 & 0 & 1 & 0 & 0 & 0 \\
1 & 0 & 0 & 0 & 0 & 1 \\
0 & 1 & 1 & 1 & 1 & 0 \\
0 & 0 & 0 & 1 & 0 & 0
\end{bmatrix}
\end{array}$$

Solution:
Let the given incidence matrix $A(G)$ be

$$A(G) = \begin{array}{c} \\ v_1 \\ v_2 \\ v_3 \\ v_4 \\ v_5 \end{array} \begin{array}{c} e_1\ e_2\ e_3\ e_4\ e_5\ e_6 \\ \begin{bmatrix} 0 & 1 & 0 & 0 & 1 & 1 \\ 1 & 0 & 1 & 0 & 0 & 0 \\ 1 & 0 & 0 & 0 & 0 & 1 \\ 0 & 1 & 1 & 1 & 1 & 0 \\ 0 & 0 & 0 & 1 & 0 & 0 \end{bmatrix} \end{array}$$

The desired graph G is shown in Fig. 6.3.

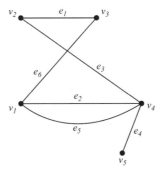

Fig. 6.3

Properties of Incidence matrix:

1. Each column of A has exactly two 1's, since every edge is incident on exactly two vertices.
2. The number of 1's in each row equals the degree of the corresponding vertex.
3. A row with all 0's represents an isolated vertex.
4. Parallel edges in a graph produce identical columns in its incidence matrix.
5. If a graph G is disconnected and consists of two components G_1 and G_2, the incidence matrix $A(G)$ of graph G can be written in a block diagonal form as

$$A(G) = \begin{bmatrix} A(G_1) & 0 \\ 0 & A(G_2) \end{bmatrix}$$

where $A(G_1)$ and $A(G_2)$ are the incidence matrices of components G_1 and G_2.

6. Permutation of any two rows or columns in an incidence matrix simply corresponds to relabeling the vertices and edges of the same graph.

Incidence Matrix of a connected Digraph:

Let G be a Digraph with n vertices, e edges. Suppose that G contains no self-loops. We define a matrix $A = (a_{ij})_{n \times e}$ whose rows correspond to the vertices and columns corresponds to the edges, as follows:

$a_{ij} = 1$, if jth edge is incident out of ith vertex

$\quad = -1$, if jth edge is incident into ith vertex

$\quad = 0$, if jth edge is neither incident out nor incident into ith vertex

Example 6.3 Find the incidence matrix of the following digraph in Fig. 6.4

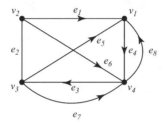

Fig. 6.4

Solution:

The incidence matrix of the given graph G in Fig. 6.4 is as follows:

$$
A(G) = \begin{array}{c} \\ v_1 \\ v_2 \\ v_3 \\ v_4 \end{array}
\begin{array}{cccccccc} e_1 & e_2 & e_3 & e_4 & e_5 & e_6 & e_7 & e_8 \\ \end{array}
\left[\begin{array}{cccccccc}
-1 & 0 & 0 & 1 & -1 & 0 & 0 & -1 \\
1 & 1 & 0 & 0 & 0 & 1 & 0 & 0 \\
0 & 1 & -1 & 0 & 1 & 0 & 1 & 0 \\
0 & 0 & 1 & -1 & 0 & -1 & -1 & 1
\end{array} \right]
$$

Theorem 6.1 *If $A(G)$ is an incidence matrix of a connected graph G with n vertices, the rank of $A(G)$ is $n - 1$.*

Proof Let G be a graph and let $A(G)$ be its incidence matrix. Now each row in $A(G)$ is a vector over $GF(2)$ in the vector space of graph G. Let the row vectors be denoted by $A_1, A_2, \ldots, A_n$. Then,

$$
A(G) = \begin{bmatrix} A_1 \\ A_2 \\ \vdots \\ A_n \end{bmatrix}
$$

Since there are exactly two 1's in every column of A, the sum of all these vectors is 0 (this being a modulo 2 sum of the corresponding entries). Thus vectors $A_1, A_2, \ldots, A_n$ are linearly dependent. Therefore, rank $A < n$.

$$\text{Hence, } \operatorname{rank} A(G) \leq n - 1 \qquad (6.1)$$

Now, consider the sum of any m of these row vectors, $m \leq n - 1$. Since G is connected, $A(G)$ cannot be partitioned in the form

$$A(G) = \begin{bmatrix} A(G_1) & 0 \\ 0 & A(G_2) \end{bmatrix}$$

such that $A(G_1)$ has m rows and $A(G_2)$ has n-m rows.

Thus, there exists no $m \times m$ submatrix of $A(G)$ for $m \leq n - 1$, such that the modulo 2 sum of these m rows is equal to zero.

As there are only two elements 0 and 1 in this field, the additions of all vectors taken m at a time for $m = 1, 2, \ldots, n - 1$ exhausts all possible linear combinations of $n - 1$ row vectors.

Thus no linear combinations of m row vectors of A, for $m \leq n - 1$, is zero.

$$\text{Therefore, } \operatorname{rank} A(G) \geq n - 1 \qquad (6.2)$$

Combining Eqs. (6.1) and (6.2), it follows that rank $A(G) = n - 1$. $\qquad \square$

Remark If G is a disconnected graph with k components, then it follows from the above theorem that rank of $A(G)$ is $n - k$.

Reduced incidence matrix: Let G be a connected graph with n vertices and m edges. Then the order of the incidence matrix $A(G)$ is $n \times m$. Now, if we remove any one row from $A(G)$, the remaining $(n - 1)$ by m submatrix is of rank $(n - 1)$. Thus, the remaining $(n - 1)$ row vectors are linearly independent. This shows that only $(n - 1)$ rows of an incidence matrix are required to specify the corresponding graph completely, because $(n - 1)$ rows contain the same information as the entire matrix. This follows from the fact that given $(n - 1)$ rows, we can construct the nth row, as each column in the matrix has exactly two 1's. Such an $(n - 1) \times m$ matrix of A is called a *reduced incidence matrix* and is denoted by A_f. The vertex corresponding to the deleted row in A_f is called the *reference vertex*. Obviously, any vertex of a connected graph can be treated as the reference vertex.

The following result gives the nature of the incidence matrix of a tree.

Theorem 6.2 *The reduced incidence matrix of a tree is nonsingular.*

Proof A tree with n vertices has $n - 1$ edges and also a tree is connected. Therefore, the reduced incidence matrix is a square matrix of order $n - 1$, with rank $n - 1$. Thus the result follows.

Now a graph G with n vertices and $n - 1$ edges which is not a tree is obviously disconnected. Therefore, the rank of the incidence matrix of G is less than $n - 1$. Hence, the $(n - 1) \times (n - 1)$ reduced incidence matrix of a graph is nonsingular if and only if the graph is a tree. $\qquad \square$

Theorem 6.3 *Let $A(G)$ be an incidence matrix of a connected graph G with n vertices. An $(n-1) \times (n-1)$ submatrix of $A(G)$ is nonsingular iff the $(n-1)$ edges corresponding to the $(n-1)$ columns of this matrix constitute a spanning tree in G.*

Proof Every square submatrix of order $n-1$ in $A(G)$ is the reduced incidence matrix of the some subgraph in G with $n-1$ edges and vice versa. A square submatrix of $A(G)$ is nonsingular iff the corresponding subgraph is a tree. The tree in this case is a spanning tree, since it contains $n-1$ edges of the n vertex graph.

Hence, $(n-1) \times (n-1)$ submatrix of $A(G)$ is nonsingular iff the $(n-1)$ edges corresponding to the $(n-1)$ columns of this matrix forms a spanning tree. □

Permutation Matrix:

A permutation matrix is a square binary matrix that has exactly one '1' in each row and column.

Theorem 6.4 *Two graphs G_1 and G_2 are isomorphic iff their adjacency matrices $X(G_1)$ and $X(G_2)$ differs only by permutations of rows and columns.*

Proof Suppose $X(G_1)$ and $X(G_2)$ are the adjacency matrices of two isomorphic graphs. Then, one of these matrices can be obtained from the other by rearranging rows and then rearranging the corresponding columns. Now rearranging rows of $X(G_1)$ is equivalent to premultiplying by a permutation matrix P yielding the product matrix $PX(G_1)$ The subsequent rearrangement of corresponding columns is equivalent to postmultiplying $PX(G_1)$ by P^{-1} (since P is nonsingular matrix). Thus $X(G_2) = PX(G_1)P^{-1}$.

Conversely, if $X(G_2)P = PX(G_1), X(G_2)$ can be obtained from $X(G_1)$ by rearranging columns and then rows, yielding that two graphs are isomorphic.

6.2.2 Adjacency Matrix

The adjacency matrix of a graph G with n vertices and no parallel edge is an n by n symmetric binary matrix $X = (x_{ij})_{n \times n}$ defined over the ring of integers such that

$$x_{ij} = 1, \text{ if there is an edge between } i\text{th and } j\text{th vertices}$$
$$= 0, \text{ if there is no edge between them}$$

To illustrate, consider the graph G as shown in Fig. 6.10.
The adjacency matrix $X(G)$ of G is given by

$$
\begin{array}{c}
\begin{array}{cccccc}
v_1 & v_2 & v_3 & v_4 & v_5 & v_6
\end{array}\\
X(G) =
\begin{array}{c}
v_1\\ v_2\\ v_3\\ v_4\\ v_5\\ v_6
\end{array}
\begin{bmatrix}
0 & 1 & 0 & 0 & 1 & 1\\
1 & 0 & 0 & 1 & 1 & 0\\
0 & 0 & 0 & 1 & 0 & 0\\
0 & 1 & 1 & 0 & 1 & 1\\
1 & 1 & 0 & 1 & 0 & 0\\
1 & 0 & 0 & 1 & 0 & 0
\end{bmatrix}
\end{array}
$$

If a graph G is disconnected and has two components g_1 and g_2 iff its adjacency matrix $X(G)$ can be partitioned as

$$
X(G) = \begin{bmatrix} X(g_1) & 0\\ 0 & X(g_2) \end{bmatrix}
$$

where $X(g_1)$ is the adjacency matrix of the component g_1 and $X(g_2)$ is the adjacency matrix of the component g_2.

Theorem 6.5 *Let X be the adjacency matrix of a simple graph G. Then the ijth entry of X^k is the number of different $v_i - v_j$ walks in G of length k.*

Proof We shall prove the result by using induction on k. The result is true for $k = 0$ and 1. For $k = 2$, the off diagonal entry in X^2, i.e., ijth entry in $X^2 (i \neq j) = $ number of different $v_i - v_j$ walks of length two.

Thus the result is true for $k = 2$.

For $k = 3$, the off diagonal entry in X^3, i.e., ijth entry in $X^3 = $ number of different $v_i - v_j$ walks of length three.

The theorem holds for $k = 1, 2, 3$.

It can be proved for any positive integer r.

Assume that, it holds for $k = r$, then evaluate the ijth entry in X^{r+1} with the help of the relation

$$
X^{r+1} = X^r . X
$$

We have, $[X^{r+1}]_{ij} = [X^r . X]_{ij} = \sum_{l=1}^{n} [X^r]_{il} [X]_{lj} = \sum_{l=1}^{n} [X^r]_{il} x_{lj}$. Now, every $v_i - v_j$ walk of length $r + 1$ consists of a $v_i - v_l$ walk of length r, followed by an edge $v_l v_j$. Since, there are $[X^r]_{il}$ such walks of length r and x_{lj} such edges for each vertex v_l, the total number of all $v_i - v_j$ walks of length $r + 1$ is $\sum_{l=1}^{n} [X^r]_{il} x_{lj}$. This completes the proof for $r + 1$ also. ☐

Theorem 6.6 *Let G be a graph with n vertices $v_1, v_2, \ldots, v_n$ and let X be the adjacency matrix of G. Let $Y = (y_{ij})_{n \times n}$ be the matrix such that*

$$Y = X + X^2 + \ldots + X^{n-1} \text{ (in the ring of integers)}$$

Then G is a connected graph iff for every entry (i,j), we have $y_{ij} \neq 0$, i.e., iff Y has no zero entries off the main diagonal.

Proof Let $x_{ij}^{(k)}$ denote the $(i,j)th$ entry of X^k, for each $k = 1, 2, \ldots, n - 1$.

Then $y_{ij} = x_{ij}^{(1)} + x_{ij}^{(2)} + \ldots + x_{ij}^{(n-1)}$

Now we know that, $x_{ij}^{(k)}$ denotes the number of distinct walks of length k from v_i to v_j and so

$$\begin{aligned} y_{ij} = &\text{(number of different } v_i - v_j \text{ walks of length 1)} \\ &+ \text{(number of different } v_i - v_j \text{ walks of length 2)}+ \ldots \\ &+ \text{(number of different } v_i - v_j \text{ walks of length}(n - 1)). \end{aligned}$$

i.e., y_{ij} is the number of different $v_i - v_j$ walks of length less than n.

Now suppose that G is connected then for every pair of vertices there is a path from v_i to v_j. Since G has only n vertices, this path goes through at most n vertices and so it has length less than n, i.e., there is at least 1 path from v_i to v_j of length less than n. Hence, $y_{ij} \neq 0$ for each i, j with $i \neq j$, as required.

Conversely, suppose that for each distinct pair i, j we have $y_{ij} \neq 0$. Then, from above there is at least one walk (of length less than n) from v_i to v_j. In particular, v_i is connected to v_j, since every $u - v$ walk contains a $u - v$ path. Thus, G is a connected graph, as required.

Example 6.4 Check whether the graph G having the following adjacency matrix X is connected or not.

$$X(G) = \begin{bmatrix} 0 & 1 & 1 & 1 & 1 & 1 \\ 1 & 0 & 1 & 0 & 0 & 0 \\ 1 & 1 & 0 & 1 & 0 & 0 \\ 1 & 0 & 1 & 0 & 1 & 0 \\ 1 & 0 & 0 & 1 & 0 & 1 \\ 1 & 0 & 0 & 0 & 1 & 0 \end{bmatrix}$$

Solution:

To check whether the given graph G is connected or not.

We need to find

$$Y = X + X^2 + X^3 + X^4 + X^5$$

$$
\text{Here, } X = \begin{bmatrix} 0 & 1 & 1 & 1 & 1 & 1 \\ 1 & 0 & 1 & 0 & 0 & 0 \\ 1 & 1 & 0 & 1 & 0 & 0 \\ 1 & 0 & 1 & 0 & 1 & 0 \\ 1 & 0 & 0 & 1 & 0 & 1 \\ 1 & 0 & 0 & 0 & 1 & 0 \end{bmatrix}, \ X^2 = \begin{bmatrix} 5 & 1 & 2 & 2 & 2 & 1 \\ 1 & 2 & 1 & 2 & 1 & 1 \\ 2 & 1 & 3 & 1 & 2 & 1 \\ 2 & 2 & 1 & 3 & 1 & 2 \\ 2 & 1 & 2 & 1 & 3 & 1 \\ 1 & 1 & 1 & 2 & 1 & 2 \end{bmatrix}
$$

Since, X^2 has all off diagonal entries nonzero.

Therefore, Y should also have all off diagonal entries nonzero.

Hence, the given graph G is connected.

Example 6.5 Show that the graph G having the following adjacency matrix X is connected.

$$
X = \begin{bmatrix} 0 & 0 & 1 & 0 & 0 \\ 0 & 0 & 0 & 1 & 0 \\ 1 & 0 & 0 & 0 & 1 \\ 0 & 1 & 0 & 0 & 1 \\ 0 & 0 & 1 & 1 & 0 \end{bmatrix}
$$

Solution:

$$
\text{Here, } Y = X + X^2 + X^3 + X^4
$$

$$
= \begin{bmatrix} 3 & 1 & 3 & 1 & 4 \\ 1 & 3 & 1 & 3 & 4 \\ 3 & 1 & 7 & 5 & 4 \\ 1 & 3 & 5 & 7 & 4 \\ 4 & 4 & 4 & 4 & 8 \end{bmatrix}
$$

Since, all off diagonal entries of Y are nonzero.

Hence, G is connected.

Verification:

The graph associated with matrix X is connected which is shown in Fig. 6.5.

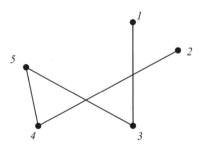

Fig. 6.5 A connected graph

Example 6.6 Check whether the graph G having the following adjacency matrix is connected or not.

$$X = \begin{bmatrix} 0 & 1 & 0 & 0 & 0 \\ 1 & 0 & 0 & 0 & 1 \\ 0 & 0 & 0 & 1 & 0 \\ 0 & 0 & 1 & 0 & 0 \\ 0 & 1 & 0 & 0 & 0 \end{bmatrix}$$

Solution:
   Here,

$$X = \begin{bmatrix} 0 & 1 & 0 & 0 & 0 \\ 1 & 0 & 0 & 0 & 1 \\ 0 & 0 & 0 & 1 & 0 \\ 0 & 0 & 1 & 0 & 0 \\ 0 & 1 & 0 & 0 & 0 \end{bmatrix}, \quad X^2 = \begin{bmatrix} 1 & 0 & 0 & 0 & 1 \\ 0 & 2 & 0 & 0 & 0 \\ 0 & 0 & 1 & 0 & 0 \\ 0 & 0 & 0 & 1 & 0 \\ 1 & 0 & 0 & 0 & 1 \end{bmatrix}$$

$$X^3 = \begin{bmatrix} 0 & 2 & 0 & 0 & 0 \\ 2 & 0 & 0 & 0 & 2 \\ 0 & 0 & 0 & 1 & 0 \\ 0 & 0 & 1 & 0 & 0 \\ 0 & 2 & 0 & 0 & 0 \end{bmatrix}, \quad X^4 = \begin{bmatrix} 2 & 0 & 0 & 0 & 2 \\ 0 & 4 & 0 & 0 & 0 \\ 0 & 0 & 1 & 0 & 0 \\ 0 & 0 & 0 & 1 & 0 \\ 2 & 0 & 0 & 0 & 2 \end{bmatrix}$$

Now, $Y = X + X^2 + X^3 + X^4$
Therefore, we have,

$$Y = \begin{bmatrix} 3 & 3 & 0 & 0 & 3 \\ 3 & 6 & 0 & 0 & 3 \\ 0 & 0 & 2 & 2 & 0 \\ 0 & 0 & 2 & 2 & 0 \\ 3 & 3 & 0 & 0 & 3 \end{bmatrix}$$

Since, in Y, there exists at least one off diagonal zero entry.
   Hence, the given graph G is not connected.
   One component of G consists of v_1, v_2, v_5 and other component consists of v_3, v_4.

6.2.3 Circuit Matrix/Cycle Matrix

Let the number of different circuits in a graph G be k and the number of edges in G be e. Then a circuit matrix $B = (b_{ij})_{k \times e}$ is a binary matrix defined as follows:

$b_{ij} = 1$, if ith circuit includes jth edge
$\quad = 0$, otherwise

It is usually denoted by $B(G)$.

Illustration: Consider the graph G_1 given in Fig. 6.6

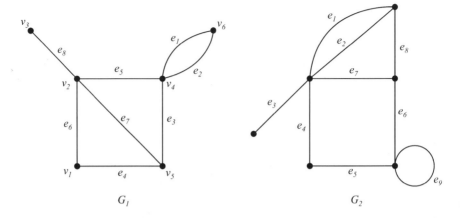

Fig. 6.6 Two graphs G_1 and G_2

The graph G_1 has four different circuits $\Gamma_1 = \{e_1, e_2\}$, $\Gamma_2 = \{e_3, e_5, e_7\}$, $\Gamma_3 = \{e_4, e_6, e_7\}$ and $\Gamma_4 = \{e_3, e_4, e_6, e_5\}$.

The circuit matrix is

$$
B(G_1) = \begin{array}{c} \\ \Gamma_1 \\ \Gamma_2 \\ \Gamma_3 \\ \Gamma_4 \end{array}
\begin{array}{c} \begin{array}{cccccccc} e_1 & e_2 & e_3 & e_4 & e_5 & e_6 & e_7 & e_8 \end{array} \\
\left[\begin{array}{cccccccc}
1 & 1 & 0 & 0 & 0 & 0 & 0 & 0 \\
0 & 0 & 1 & 0 & 1 & 0 & 1 & 0 \\
0 & 0 & 0 & 1 & 0 & 1 & 1 & 0 \\
0 & 0 & 1 & 1 & 1 & 1 & 0 & 0
\end{array} \right] \end{array}
$$

The graph G_2 of Fig. 6.6 has seven different circuits, namely,
$\Gamma_1 = \{e_1, e_2\}$, $\Gamma_2 = \{e_2, e_7, e_8\}$, $\Gamma_3 = \{e_1, e_7, e_8\}$, $\Gamma_4 = \{e_4, e_5, e_6, e_7\}$, $\Gamma_5 = \{e_2, e_4, e_5, e_6, e_8\}$, $\Gamma_6 = \{e_1, e_4, e_5, e_6, e_8\}$ and $\Gamma_7 = \{e_9\}$.

The cycle matrix is given by

$$
B(G_2) = \begin{array}{c} \\ \Gamma_1 \\ \Gamma_2 \\ \Gamma_3 \\ \Gamma_4 \\ \Gamma_5 \\ \Gamma_6 \\ \Gamma_7 \end{array}
\begin{array}{c} \begin{array}{ccccccccc} e_1 & e_2 & e_3 & e_4 & e_5 & e_6 & e_7 & e_8 & e_9 \end{array} \\
\left[\begin{array}{ccccccccc}
1 & 1 & 0 & 0 & 0 & 0 & 0 & 0 & 0 \\
0 & 1 & 0 & 0 & 0 & 0 & 1 & 1 & 0 \\
1 & 0 & 0 & 0 & 0 & 0 & 1 & 1 & 0 \\
0 & 0 & 0 & 1 & 1 & 1 & 1 & 0 & 0 \\
0 & 1 & 0 & 1 & 1 & 1 & 0 & 1 & 0 \\
1 & 0 & 0 & 1 & 1 & 1 & 0 & 1 & 0 \\
0 & 0 & 0 & 0 & 0 & 0 & 0 & 0 & 1
\end{array} \right] \end{array}
$$

We have the following observations regarding the circuit matrix $B(G)$ of a graph G,

1. A column of all zeros corresponds to a non-cycle edge, that is, an edge which does not belong to any cycle.
2. Each row of $B(G)$ is a cycle vector.
3. A cycle matrix has the property of representing a self-loop and the corresponding row has a single 1.
4. The number of 1's in a row is equal to the number of edges in the corresponding circuit.
5. If the graph G is separable (or disconnected) and consists of two blocks (or components) H_1 and H_2, then the circuit matrix $B(G)$ can be written in a block-diagonal form as

$$B(G) = \begin{bmatrix} B(H_1) & 0 \\ 0 & B(H_2) \end{bmatrix},$$

where $B(H_1)$ and $B(H_2)$ are the circuit matrices of H_1 and H_2. This follows from the fact that circuit in H_1 have no edges belonging to H_2 and vice versa.

6. Permutation of any two rows or columns in a circuit matrix corresponds to relabeling the circuits and the edges.
7. Two graphs G_1 and G_2 are 2-isomorphic if and only if they have circuit correspondence. Thus two graphs G_1 and G_2 have the same circuit matrix if and only if G_1 and G_2 are 2-isomorphic.

For example, the two graphs given in Fig. 6.7 have the same circuit matrix. They are 2-isomorphic, but are not isomorphic.

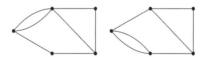

Fig. 6.7 2-isomorphic graphs

The following result relates the incidence and circuit matrix of a graph without self-loops.

Theorem 6.7 *If G is a graph without self-loops, with incidence matrix A and circuit matrix B whose columns are arranged using the same order of edges, then every row of B is orthogonal to every row of A, that is $AB^T = BA^T \equiv 0 \ (mod\ 2)$, where A^T and B^T are the transposes of A and B, respectively.*

Proof Let G be a graph without self-loops and let A and B respectively, be the incidence and circuit matrix of G.

We know that in G for any vertex v_i and for any cycle Γ_j, either $v_i \in \Gamma_j$ or $v_i \notin \Gamma_j$. In case $v_i \notin \Gamma_j$, then there is no edge of Γ_j which is incident on v_i and if $v_i \in \Gamma_j$, then there are exactly two edges of Γ_j which are incident on v_i.

Now, consider the ith row of A and the jth row of B (which is the jth column of B^T). Since the edges are arranged in the same order, the rth entries in these two rows are both nonzero if and only if the edge e_r is incident on the ith vertex v_i and is also in the jth cycle Γ_j.

We have $[AB^T]_{ij} = \sum [A]_{ir}[B^T]_{rj} = \sum [A]_{ir}[B]_{jr} = \sum a_{ir}b_{jr}$

For each e_r of G, we have one of the following cases.

i. e_r is incident on v_i and $e_r \notin \Gamma_j$. Here $a_{ir} = 1$, $b_{jr} = 0$.
ii. e_r is not incident on v_i and $e_r \in \Gamma_j$. In this case, $a_{ir} = 0$, $b_{jr} = 1$.
iii. e_r is not incident on v_i and $e_r \notin \Gamma_j$, so that $a_{ir} = 0, b_{jr} = 0$.

All these cases imply that the ith vertex v_i is not in the jth cycle Γ_j and we have $[AB^T]_{ij} = 0 \equiv 0 \,(\mathrm{mod}\,2)$.

iv. e_r is incident on v_i and $e_r \in \Gamma_j$.

Here we have exactly two edges, say e_r and e_t incident on v_i so that $a_{ir} = 1, a_{it} = 1$, $b_{jr} = 1, b_{jt} = 1$. Therefore, $[AB^T]_{ij} = \sum a_{ir}b_{jr} = 1 + 1 \equiv 0 \,(\mathrm{mod}\,2)$.
□

We illustrate the above theorem with the following example in Fig. 6.8

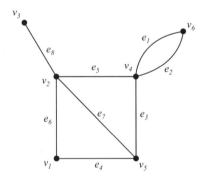

Fig. 6.8

Clearly,

$$AB^T = \begin{bmatrix} 0 & 0 & 0 & 1 & 0 & 1 & 0 & 0 \\ 0 & 0 & 0 & 0 & 1 & 1 & 1 & 1 \\ 0 & 0 & 0 & 0 & 0 & 0 & 0 & 1 \\ 1 & 1 & 1 & 0 & 1 & 0 & 0 & 0 \\ 0 & 0 & 1 & 1 & 0 & 0 & 1 & 0 \\ 1 & 1 & 0 & 0 & 0 & 0 & 0 & 0 \end{bmatrix} \begin{bmatrix} 1 & 0 & 0 & 0 \\ 1 & 0 & 0 & 0 \\ 0 & 1 & 0 & 1 \\ 0 & 0 & 1 & 1 \\ 0 & 1 & 0 & 1 \\ 0 & 0 & 1 & 1 \\ 0 & 1 & 1 & 0 \\ 0 & 0 & 0 & 0 \end{bmatrix}$$

$$= \begin{bmatrix} 0 & 0 & 2 & 2 \\ 0 & 2 & 2 & 2 \\ 0 & 0 & 0 & 0 \\ 2 & 2 & 0 & 2 \\ 0 & 2 & 2 & 2 \\ 2 & 0 & 0 & 0 \end{bmatrix} \equiv 0 \,(mod\,2)$$

Again,

$$BA^T = \begin{bmatrix} 1 & 1 & 0 & 0 & 0 & 0 & 0 & 0 \\ 0 & 0 & 1 & 0 & 1 & 0 & 1 & 0 \\ 0 & 0 & 0 & 1 & 0 & 1 & 1 & 0 \\ 0 & 0 & 1 & 1 & 1 & 1 & 0 & 0 \end{bmatrix} \begin{bmatrix} 0 & 0 & 0 & 1 & 0 & 1 \\ 0 & 0 & 0 & 1 & 0 & 1 \\ 0 & 0 & 0 & 1 & 1 & 0 \\ 1 & 0 & 0 & 0 & 1 & 0 \\ 0 & 1 & 0 & 1 & 0 & 0 \\ 1 & 1 & 0 & 0 & 0 & 0 \\ 0 & 1 & 0 & 0 & 1 & 0 \\ 0 & 1 & 1 & 0 & 0 & 0 \end{bmatrix}$$

$$= \begin{bmatrix} 0 & 0 & 0 & 2 & 0 & 2 \\ 0 & 2 & 0 & 2 & 2 & 0 \\ 2 & 2 & 0 & 0 & 2 & 0 \\ 2 & 2 & 0 & 2 & 2 & 0 \end{bmatrix} \equiv 0 \,(mod\,2)$$

6.2.3.1 Fundamental Circuit Matrix

A submatrix (of a circuit matrix) in which all rows correspond to a set of fundamental circuits is called a fundamental circuit matrix B_f.

The graph G of Fig. 6.9 has three different fundamental circuits with regard to the chord e_2, e_3 and e_6 viz. $\Gamma_1 = \{e_1, e_2, e_4, e_7\}$, $\Gamma_2 = \{e_3, e_4, e_7\}$ and $\Gamma_3 = \{e_5, e_6, e_7\}$

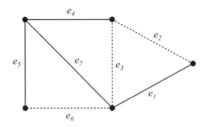

Fig. 6.9

The fundamental circuit matrix B_f of Fig. 6.9 is as follows

$$
B_f = \begin{array}{c} \\ \Gamma_1 \\ \Gamma_2 \\ \Gamma_3 \end{array}
\begin{array}{ccccccc}
e_2 & e_3 & e_6 & e_1 & e_4 & e_5 & e_7 \\
\end{array}
\left[
\begin{array}{ccc|cccc}
1 & 0 & 0 & 1 & 1 & 0 & 1 \\
0 & 1 & 0 & 0 & 1 & 0 & 1 \\
0 & 0 & 1 & 0 & 0 & 1 & 1
\end{array}
\right]
$$

A matrix B_f thus arranged can be written as

$$B_f = \begin{bmatrix} I_\mu & | & B_t \end{bmatrix} \tag{6.3}$$

where I_μ is an identity matrix of order $\mu = e - n + 1$ and B_t is the remaining $\mu \times (n - 1)$ submatrix, corresponding to the branches of the spanning tree.

From Eq. (6.3), we have

$$\text{Rank of } B_f = e - n + 1.$$

Since, B_f is a submatrix of the circuit matrix B,

$$\text{Rank of } B \geq e - n + 1 \tag{6.4}$$

Theorem 6.8 *If B is a circuit matrix of a connected graph G with e edges and n vertices*

$$\text{Rank of } B = e - n + 1$$

Proof The set of all circuit vertices in W_G forms a subspace W_Γ. This subspace W_Γ is called the circuit subspace. Every row in circuit matrix B is a circuit vector in W_Γ.

Rank of matrix B = Number of linearly independent rows in B.

But the number of independent rows in $B \leq$ number of linearly independent vectors in W_Γ and consequently, the number of linearly independent vectors in $W_\Gamma \leq$ (dimension of $W_\Gamma)^1 = \mu = e - n + 1$. Since, the set of circuit vectors, corresponding to the set of fundamental circuits with regard to any spanning tree, forms a basis for the circuit subspace W_Γ and the number of circuit vectors (including zero vector **0**) in W_Γ is 2^μ.

$$\text{Therefore, Rank of } B \leq e - n + 1. \tag{6.5}$$

Again, in Eq. (6.4) we have just shown that

$$\text{Rank of } B \geq e - n + 1.$$

Hence, combining Eqs. (6.4) and (6.5), we have

$$\text{Rank of } B = e - n + 1. \qquad \square$$

Theorem 6.9 *The determinant of every square submatrix of the incidence matrix A of a digraph is* 1, -1 *or* 0.

Proof Consider a k by k submatrix M of A. If M has any column or row consisting of all zeros, det M is clearly zero. Also, det $M = 0$ if every column of M contains the two non-zero entries 1 and -1. (since, in this case, row sum of M is zero). Now if det $M \neq 0$ (i.e. M is nonsingular). Then the sum of entries in each column of M cannot be zero. Therefore, M must have a column in which there is a single non zero element that is either $+1$ or -1.

Let this single element be in the (i,j) th position in M. Thus

$$\det M = \pm \det M_{ij}$$

where M_{ij} is the submatrix of M with its ith row and jth column deleted. The $(k-1)$ by $(k-1)$ submatrix M_{ij} is also nonsingular (because M is nonsingular). Therefore, it must also have at least one column with a single non-zero entry say in the (m,n)th position in M_{ij}. Expanding det M_{ij} about this element in the (m, n)th position, det $M_{ij} = \pm$ (determinant of a nonsingular $(k-2)$ by $(k-2)$ submatrix of M). Repeated application of this procedure yields

$$\det M = \pm 1$$

Hence, the theorem is proved. $\qquad \square$

---

[1] A vector space is n-dimensional if the maximum number of linearly independent vectors in the space is n. If a vector space V has a basis $\beta = \{b_1, b_2, \ldots, b_n\}$, then any set in V containing more than n vectors must be linearly dependent.

Exercises:

1. Determine the graph whose incidence matrix is

$$
\begin{array}{c c c c c c}
 & e_1 & e_2 & e_3 & e_4 & e_5 \\
\begin{array}{c} v_1 \\ v_2 \\ v_3 \\ v_4 \\ v_5 \\ v_6 \end{array} &
\left[\begin{array}{c c c c c}
0 & 0 & 1 & -1 & 1 \\
-1 & 1 & 0 & 0 & 0 \\
0 & 0 & 0 & 0 & 0 \\
1 & 0 & 0 & 0 & -1 \\
0 & 1 & 0 & 0 & 0 \\
0 & 0 & -1 & 1 & 0
\end{array}\right]
\end{array}
$$

2. Find the graph whose incidence matrix is

$$
\begin{array}{c c c c c c}
 & e_1 & e_2 & e_3 & e_4 & e_5 \\
\begin{array}{c} v_1 \\ v_2 \\ v_3 \\ v_4 \\ v_5 \end{array} &
\left[\begin{array}{c c c c c}
1 & 1 & 1 & 0 & 0 \\
0 & 1 & 0 & 0 & 0 \\
1 & 0 & 1 & 0 & 0 \\
0 & 0 & 0 & 1 & 1 \\
0 & 0 & 0 & 1 & 1
\end{array}\right]
\end{array}
$$

3. Determine the incidence matrix of the following graph (Fig. 6.10)

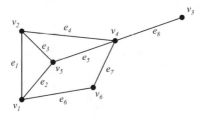

Fig. 6.10

4. Using the incidence matrices, find whether the two graphs G_1 and G_2 are isomorphic or not (Fig. 6.11).

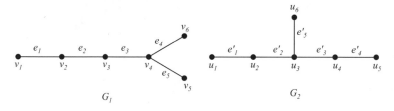

Fig. 6.11 Two graphs G_1 and G_2

5. Show that, the following two graphs are not isomorphic but they are 2-isomorphic (Fig. 6.12).

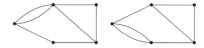

Fig. 6.12 Two non-isomorphic graphs

6. Show that, the determinant of every square submatrix of the incidence matrix A, of the following digraph in Fig. 6.13, is either 1, -1 or 0.

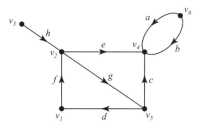

Fig. 6.13 A digraph

7. Prove that, a graph is bipartite if and only if for all odd k, every diagonal entry of A^k is zero.
8. Check whether the graph G having the following adjacency matrix is connected or not

$$X(G) = \begin{bmatrix} 0 & 1 & 1 & 0 & 0 \\ 0 & 0 & 0 & 1 & 0 \\ 1 & 0 & 0 & 0 & 1 \\ 0 & 1 & 0 & 0 & 1 \\ 0 & 0 & 1 & 1 & 0 \end{bmatrix}$$

9. Establish the isomorphism of the two graphs given in the following figures by considering their adjacency matrices (Fig. 6.14)

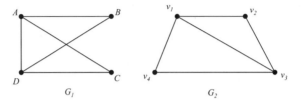

Fig. 6.14 Two graphs G_1 and G_2

Chapter 7
Cut Sets and Cut Vertices

In this chapter, we find a type of subgraph of a graph G where removal from G separates some vertices from others in G. This type of subgraph is known as cut set of G. Cut set has a great application in communication and transportation networks.

7.1 Cut Sets and Fundamental Cut Sets

7.1.1 Cut Sets

Fig. 7.1 A cut set $\{e_2, e_5, e_9, e_{10}\}$ of the graph

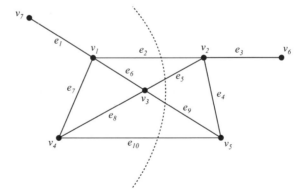

In a connected graph G, the set of edges is said to be a *cut set* of G if removal of the set from G leaves G disconnected but no proper subsets of this set does not do so.

In the graph shown in Fig. 7.1, the set of edges $\{e_2, e_5, e_9, e_{10}\}$ is a cut set of the graph. In Fig. 7.1, it is represented by a dotted curve. It can be noted that the edge set $\{e_2, e_5, e_4\}$ is also a cut set of the graph. $\{e_1\}$ is a cut set containing only one edge. Removal of the set $\{e_2, e_5, e_4, e_9, e_{10}\}$ disconnects the graph but it is not cut set because its proper subset $\{e_2, e_5, e_4\}$ is a cut set.

S. Saha Ray, *Graph Theory with Algorithms and its Applications*, DOI: 10.1007/978-81-322-0750-4_7, © Springer India 2013

Corollary

(1) *Every edge of a tree is a cut set of the tree.*

(2) *A cut set is a subgraph of G.*

7.1.2 Fundamental Cut Set (or Basic Cut Set)

Let T be a spanning tree of a connected graph G. A cut set S of G containing exactly one branch of T is called a *Fundamental cut set* of G with regard to T.

Fig. 7.2 $\{BD, CD, EF\}$ is a cut set but not a fundamental cut set

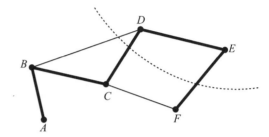

In Fig. 7.2, $T = \{AB, BC, CD, DE, EF\}$ is a spanning tree (shown by bold lines). The set of edges $\{BD, BC\}$ is a cut set containing one branch BC of T. So, $\{BD, BC\}$ is a Fundamental cut set of G w.r.t T. In the same graph the set $\{BD, CD, EF\}$ (as shown by dotted curve in Fig. 7.2) is a cut set but not a Fundamental Cut set with regard to T because it contains two edges of T.

7.2 Cut Vertices

A vertex v of a connected graph G is said to be a *cut vertex* if its deletion from G (together with the edges incident to it) disconnects the graph.

Fig. 7.3 **a** Vertex A is a cut vertex of the graph, **b** Disconnected graph after the removal of vertex A

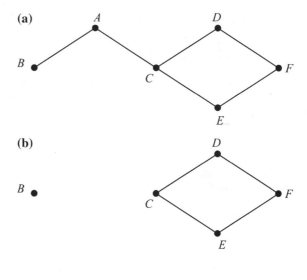

Fig. 7.4 A disconnected graph with two connected components

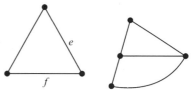

In the graph G shown in Fig. 7.3a removal of the vertex A (together with the edges incident to it) leaves a disconnected graph shown in Fig. 7.3b. So, A is a cut vertex, we note that C is also a cut vertex. But F is not a cut vertex of this graph.

Corollary

(1) *Every vertex (with degree greater than one) of a tree is cut vertex.*
(2) *A graph may have no cut vertex at all. For example, the graph in* Fig. 7.4 *has no cut vertex. Another examples are* K_2, K_3, K_4, *etc. They have no cut vertex.*

7.2.1 Cut Set with respect to a Pair of Vertices

If a cut set puts two vertices v_1 and v_2 into two different components. Then, it is called a cut set with regard to v_1 and v_2.

In the graph shown in Fig. 7.3a, $\{AC\}$ is a cut set with regard to the vertices B and E. This is not a cut set with regard to A and B.

7.3 Separable Graph and its Block

7.3.1 Separable Graph

A connected graph (or a connected component of a graph) is said to be *separable* if it has a cut vertex.

On the other hand, a connected graph (or a connected component of a graph) which is not separable is called *non-separable* graph.

The graph in Fig. 7.3a is separable. On the other hand, each of the two connected components of the graph in Fig. 7.4 is non-separable. Again each of the two components in Fig. 7.3b is nonseparable.

7.3.2 Block

A separable graph consists of two or more non-separable subgraphs. Each of these non-separable subgraphs is called a *Block*. The graph in Fig. 7.3a has the following Blocks shown in Fig. 7.5.

Fig. 7.5 Three Blocks of the graph in Fig. 7.3a

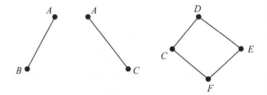

It can be noted that each of these blocks has no cut vertex.

But between the following two subgraphs of Fig. 7.3a shown in Fig. 7.6, first one is not a block because it is further separable as *A* is cut vertex of it.

Fig. 7.6 The first subgraph of Fig.7.3a is not a Block whereas the second subgraph is a Block

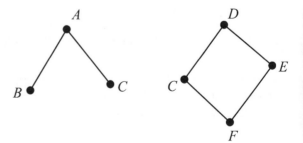

7.4 Edge Connectivity and Vertex Connectivity

7.4.1 Edge Connectivity of a Graph

Let G be a graph (may be disconnected) having k components. The minimum number of edges whose deletion from G increases the number of components of G is called *edge connectivity* of G. It is denoted by $\lambda(G)$.

In Fig. 7.4, a graph having two components is shown. We see that if one edge is deleted from the graph, the number of its components still remains 2. But if two particular edges, say e and f are deleted then number of components becomes 3. So, the edge connectivity of the graph is 2.

Corollary

(1) *The number of edges in the smallest cut set of a graph is its edge connectivity.*
(2) *The edge connectivity of a tree is 1.*

7.4.2 Vertex Connectivity of a Graph

Let G be a graph (may be disconnected). The minimum number of vertices (together with the edges incident to it) whose deletion from G increase the number of components of G is called *vertex connectivity* of G. It is denoted by $\kappa(G)$.

k-connected and k-edge connected: A graph G is *k-connected* if $\kappa(G) = k$, and G is *k-edge connected* if $\lambda(G) = k$

For Example, the vertex connectivity of the graph in Fig. 7.7 is 2.

Fig. 7.7 The vertex connectivity of the graph is 2

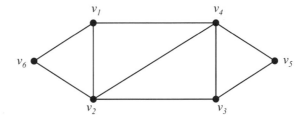

Corollary

(1) *The vertex connectivity of tree is 1.*
(2) *The vertex connectivity of a connected separable graph is 1.*

Theorem 7.1 *Every cut set in a connected graph contains at least one branch of every spanning tree of the graph.*

Proof Let, S be a cut set of G. Let T be a spanning tree of G. Suppose that, S does not contain any branch of T. Then all edges of T are present in $G - S$. It means that $G - S$ is connected graph. It implies that S is not a cut set. Hence a cut set must contain at least one branch of a spanning tree of G. ☐

For example, in Fig. 7.8, $\{a, c\}$ is not a cut set, so it should contain one branch of T to become a cut set.

Fig. 7.8 A tree G and its spanning tree T

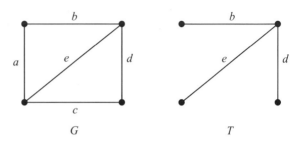

Theorem 7.2 *A vertex v in a connected graph G is a cut vertex if and only if there exists two vertices a and b distinct from v in G such that every path connecting a and b passes through v.*

Proof If v is a cut vertex of G, $G - v$ is a disconnected graph. Let us select two vertices a and b in two different components of $G - v$. Then there exists no path from a to b in $G - v$. Since, G is connected graph there exists a path P from a to b in G. If the path does not contain the vertex v, then removal of v from G will not disconnect the vertices a and b, which is a contradiction to the fact that a and b lies in two different components of $G - v$. Hence every path between a and b passes through v.

Conversely, if every path from a to b contains the vertex v then removal of v from G disconnects a and b. Hence, a and b lies in different components of G which implies that $G - v$ is disconnected graph. Therefore, v is cut vertex of G. ☐

Theorem 7.3 *The edge connectivity of a graph ≤ the smallest degree of all vertices of the graph.*

Proof Let v_k be the vertex with smallest degree in G. Let $d(v_k)$ be the degree of v_k. Vertex v_k can be separated from G by removing the $d(v_k)$ edges incident on vertex v_k. Therefore, removal of $d(v_k)$ edges disconnects the graph. Hence, the edge connectivity of a graph cannot exceed the smallest degree of all vertices of the graph. ☐

Theorem 7.4 *In any graph, the vertex connectivity ≤ the edge connectivity.*

Proof Let λ denote the edge connectivity of G. Therefore, there exists a cut set S in G containing λ edges.

Then, S partitions the vertices of G into two subsets V_1 and V_2 such that every edge in S joins a vertex in V_1 to a vertex in V_2. By removing at most λ vertices from V_1 or V_2 on which the edges of S are incident, we will be able to remove S (together with all other edges incident on these vertices) from G. Thus, removal of at most λ vertices from G will disconnect the graph. Hence, the vertex connectivity is less than or equal to λ.   □

Corollary *Every cut set in a non-separable graph with more than two vertices contains at least two edges.*

Proof A graph is nonseparable if its vertex connectivity is at least two. In view of Theorem 7.4, edge connectivity ≥ vertex connectivity. Hence, edge connectivity of a non-separable graph is at least two which is possible if the graph has at least two edges.   □

Theorem 7.5 *The maximum vertex connectivity of a connected graph with n vertices and e edges $(e \geq n - 1)$ is the integral part of the number $2e/n$, i.e., $\lfloor 2e/n \rfloor$. (The floor function **floor(x)** = $\lfloor x \rfloor$ is the largest integer not greater than x)*

Proof We know that every edge in G contributes two degrees. Thus the sum of degrees of all the vertices is $2e$. Since, this sum $2e$ is divided among n vertices, therefore, there must be at least one vertex in G whose degree is less than or equal to the number $2e/n$.

Therefore, using Theorem 7.3, the edge connectivity of $G \leq 2e/n$.

Consequently, it follows from Theorem 7.4, vertex connectivity ≤ edge connectivity $\leq 2e/n$.

Hence, maximum vertex connectivity possible is $\lfloor 2e/n \rfloor$.   □

Theorem 7.6 (Whitney's Inequality) *For any graph G, $\kappa(G) \leq \lambda(G) \leq \delta(G)$ i.e. vertex connectivity ≤ the edge connectivity ≤ the minimum degree of the graph G*

Proof We shall first prove $\lambda(G) \leq \delta(G)$.

If G has no edges, then $\lambda = 0$ and $\delta = 0$. If G has edges, then we get a disconnected graph, when all edges incident with a vertex of minimum degree are removed. Thus, in either case, $\lambda(G) \leq \delta(G)$.

Now, from Theorem 7.4, it follows that $\kappa(G) \leq \lambda(G)$. Hence, it is proved.   □

Example 7.1   Find the edge connectivity, vertex connectivity and minimum degree of the following graph in Fig. 7.9

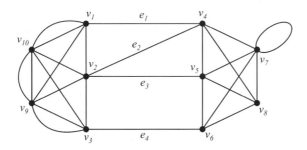

Fig. 7.9

Solution:
The vertex connectivity of the given graph is three because removal of v_1, v_2, v_3 or v_4, v_5, v_6 disconnects the graph.

The edge connectivity of this graph is four. $S = \{e_1, e_2, e_3, e_4\}$ is one such cut set. It can be observed that the degree of each vertex is at least four.

Therefore, $\kappa(G) = 3$, $\lambda(G) = 4$ *and* $\delta(G) = 4$.

Example 7.2   Find the edge connectivity, vertex connectivity and minimum degree of the following graph in Fig. 7.10.

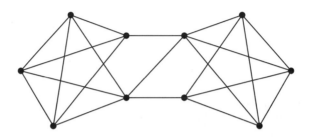

Fig. 7.10

Solution:
The vertex connectivity of the given graph is two because removal of at least two vertices are required to disconnect the graph.

The edge connectivity of this graph is three because removal of at least three edges are required to disconnect the graph. It can be observed that the degree of each vertex is at least four.

Therefore, $\kappa(G) = 2$, $\lambda(G) = 3$ *and* $\delta(G) = 4$.

Example 7.3   Show that, the edge connectivity $\lambda(G)$, vertex connectivity $\kappa(G)$, and minimum degree $\delta(G)$of the following graph in Fig. 7.11 are equal. Is the given graph separable?

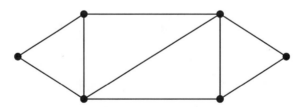

Fig. 7.11

Solution:
The vertex connectivity of the given graph is two because removal of at least two vertices are required to disconnect the graph. So, $\kappa(G) = 2$

The edge connectivity of this graph is two because removal of at least two edges are required to disconnect the graph. Therefore, $\lambda(G) = 2$

It can be observed that the degree of each vertex is at least two.
Therefore, $\kappa(G) = \lambda(G) = \delta(G) = 2$.
Moreover, the given graph in Fig. 7.11 is nonseparable, since it has no cut vertex.

Example 7.4   Find the fundamental cut sets of the graph in Fig. 7.1
Solution:
The Fig. 7.12 shows the spanning tree obtained by DFS.

Fig. 7.12 A spanning tree obtained by DFS

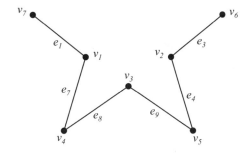

From the simple graph Fig. 7.1, we see that there are $n - 1 = 7 - 1 = 6$ fundamental cut sets with regard to the branches e_1, e_3, e_4, e_7, e_8 and e_9 of the spanning tree in Fig. 7.12.

Fundamental cut sets	Corresponding branch
$\{e_2, e_6, e_7\}$	e_7
$\{e_2, e_5, e_4\}$	e_4
$\{e_2, e_6, e_8, e_{10}\}$	e_8
$\{e_2, e_5, e_9, e_{10}\}$	e_9
$\{e_1\}$	e_1
$\{e_3\}$	e_3

Exercises:

1. Prove that a vertex v of a tree T is a cut vertex if and only if $d(v) > 1$.
2. Let T be a tree with at least three vertices. Prove that there is a cut vertex v of T such that every vertex adjacent to v, except for possibly one, has degree 1.
3. Let v be a cut vertex of the simple connected graph G. Prove that v is not a cut vertex of its complement $\overline{G}$.
4. Let G be a simple connected graph with at least two vertices and let v be a vertex in G of smallest possible degree, say k.
   (a) Prove that $\kappa(G) \leq k$ where $\kappa(G)$ is called vertex connectivity of G. It is the smallest number of vertices in G whose deletion from G leaves either a disconnected graph or K_1.
   (b) Prove that $\kappa(G) \leq 2e/n$, where e is the number of edges and n is the number of vertices in G.

5. Let G be a Hamiltonian graph. Show that G does not have a cut vertex.
6. Find the edge connectivity and vertex connectivity of the following graph in Fig. 7.13.

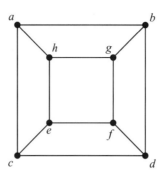

Fig. 7.13

Chapter 8
Coloring

Coloring of a graph

Proper coloring: A proper coloring or coloring of a graph G assigns colors usually denoted by 1, 2, 3, … etc., to the vertices of G, one color per vertex, so that the adjacent vertices are assigned different colors. For example, Fig. 8.1 shows proper coloring of a graph.

8.1 Properly Colored Graph

A graph in which every vertex has been assigned a color according to a proper coloring is called a properly colored graph.

Fig. 8.1 A properly colored graph

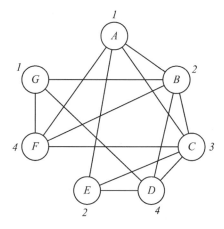

k-Colorable: A k-coloring of G is a coloring which consists of k different colors and in this case, the graph G is said to be k-colorable.

S. Saha Ray, *Graph Theory with Algorithms and its Applications*,
DOI: 10.1007/978-81-322-0750-4_8, © Springer India 2013

8.2 Chromatic Number

The minimum number n, for which there is an n-coloring of the graph G, is called chromatic number or chromatic index of the graph G. It is denoted by $\chi(G)$.

If $\chi(G) = k$, then the corresponding graph G is called k-chromatic. For example, the graph in Fig. 8.1 is 4-chromatic.

Theorem 8.1 *Let G be a non-empty graph. Then $\chi(G) = 2$, i.e.,. 2-chromatic if and only if G is bipartite.*

Proof Let G be a bipartite graph with bipartition $V = X \cup Y$.

Now, we assign color 1 to all the vertices in X and assign color 2 to all the vertices in Y. Then it gives a 2-coloring for G and so, since G is nonempty, $\chi(G) = 2$.

Conversely,

suppose that $\chi(G) = 2$, i.e., the graph G is 2-chromatic. Then, G has a 2-coloring.

To show: The graph G is bipartite.

Let X be the set of all vertices with color 1 and Y be the set of all vertices with color 2.

Since, the graph G is properly colored, no two vertices in X are adjacent to each other. Similarly, no two vertices in Y are adjacent to each other.

So, since G is nonempty, every edge of G has one end vertex in X and another end vertex in Y. Therefore, G is a bipartite graph with bipartition $V = X \cup Y$. $\square$

Theorem 8.2 *Let G be a graph then $\chi(G) \geq 3$ if and only if G has an odd cycle.*

Proof Let G be a non-empty graph with at least two vertices, then G is bipartite if and only if it has no odd cycle. Then from the previous Theorem 8.1, $\chi(G) = 2$ if and only if G has no odd cycle.

Now, if the graph G has an odd cycle then the chromatic number of G should be greater than 2, since the graph has at least two vertices the chromatic number of G cannot be less than 2 , i.e., $\chi(G) \neq 1$. Moreover, we would require at least three colors just for that odd cycle in G. So, $\chi(G) \geq 3$ if and only if the graph G has an odd cycle. $\square$

Theorem 8.3 *A graph with at least one edge is 2-chromatic if and only if it has no odd cycles.*

Proof It follows from Theorem 8.1, since we know that if G has a non-empty graph with at least two vertices, then G is bipartite if and only if G has no odd cycles. $\square$

Theorem 8.4 *Every tree with two or more vertices is 2-chromatic.*

Proof Consider the vertex v be the root of the tree T as shown in Fig. 8.2.

Fig. 8.2 A Properly colored
tree

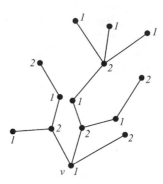

Now color the root vertex v with 1 then color all the vertices which are adjacent to the root vertex v with color 2. Again color the vertices adjacent to these vertices using color 1. Continue this process until all the vertices in T has been properly colored.

From the tree T, we can see that all the vertices at odd distances from the root vertex v have color 2, while v and the vertices at even distances from v have color 1.

Now consider any path in the tree T, the vertices along that path are alternating colored. Since there is one and only one path between any two vertices in a tree, no two adjacent vertices have the same color. Thus, T has been properly colored with two colors, viz. $\chi(T) = 2$. So, T is 2-chromatic. Hence, it is proved.   □

8.3 Chromatic Polynomial

In general, a given graph G with n vertices can be properly colored in many different ways using a sufficiently large number of colors. This property of a graph can be expressed elegantly by means of a polynomial. This polynomial is called the *Chromatic Polynomial of G*.

Definition: The chromatic polynomial gives the value which indicates the number of different ways of properly coloring the graph G with n-vertices using at most λ colors (λ or fewer numbers of colors). It is usually denoted by $\chi_G(\lambda)$ or $\chi(G; \lambda)$.

8.3.1 Chromatic Number Obtained by Chromatic Polynomial

The smallest positive integer value of λ, such that $\chi_G(\lambda)$ is not equal to zero, is the Chromatic Number of G.

So, $\chi(G) = \min \{ \lambda \in \mathbb{Z}^+ | \chi_G(\lambda) \neq 0 \}$.

For example, if the graph be K_2 and if there are λ colors available, one of the vertices can be colored in λ different ways and the other can be colored in $(\lambda - 1)$ different ways. So, $\chi_{K_2}(\lambda) = \lambda(\lambda - 1)$. The smallest positive integer value of λ such that $\chi_{K_2}(\lambda) = \lambda(\lambda - 1) \neq 0$ is 2, which is the Chromatic Number of K_2.

More generally,

$$\chi_{K_n}(\lambda) = \lambda(\lambda - 1)(\lambda - 2) \cdots (\lambda - n + 1)$$

8.3.2 Chromatic Polynomial of a Graph G

Suppose that, there are $f(r)$ different ways of partitioning the vertex set of a graph G into r independent (A set of vertices in a graph G is independent if no two of them are adjacent.) non-empty subsets (In other words, it is the number of different ways of partitioning the vertex set of G so that the vertices of G can be properly colored using r colors). Then for each such partition, the number of different ways of properly coloring the vertices of G is $\lambda^{(r)}$, where $\lambda^{(r)} = \lambda(\lambda - 1)(\lambda - 2) \cdots$ $(\lambda - r + 1)$ (which is called Factorial Function).

[If n is any positive integer, then the factorial nth power of λ is denoted by $[\lambda]^n$ or $\lambda^{(n)}$ and is defined by $\lambda^{(n)} = \lambda(\lambda - 1)(\lambda - 2) \cdots (\lambda - n + 1)$. In particular, $\lambda^{(0)} = 1$ and $\lambda^{(1)} = \lambda$]

Consequently,

$$\chi_G(\lambda) = \sum_r f(r)\lambda^{(r)}$$

If the order of G be n , i.e., G has n vertices, $f(r) = 0$ whenever $r > n$.

Thus, $\chi_G(\lambda)$ is a polynomial in λ, which is known as Chromatic Polynomial of degree n with integer co-efficients. In this polynomial, the co-efficient of the leading term λ^n is 1, since $f(n) = 1$, i.e., there is only one way of partitioning the vertex set of n vertices into n non-empty independent subsets. Moreover, $f(0) = 0$; therefore the constant term in the Chromatic polynomial is 0. It can be observed that the co-efficients in the polynomial alternate in sign.

Example 8.1 Find the chromatic polynomial of the graph in Fig. 8.3.

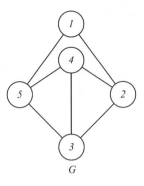

G

Fig. 8.3

Solution:
Let $f(r)$ be the number of different ways of partitioning the vertex set $V = \{1, 2, 3, 4, 5\}$ into r independent subsets. Hence, there is at least one edge in the graph. So, $f(1) = 0$. It is not possible to partition V into two independent subsets. So, $f(2) = 0$. In fact, the vertex set V cannot be partitioned into one or two independent subsets so that the vertices of G can be properly colored using one or two colors, respectively.

There are two ways of partitioning V into three independent subsets, viz. $\{\{3\}, \{1, 4\}, \{2, 5\}\}$ and $\{\{4\}, \{1, 3\}, \{2, 5\}\}$. So, $f(3) = 2$. Consequently, there are two ways of partitioning V so that the vertices of G can be properly colored using precisely three colors.

Similarly, there are three ways of partitioning V into four independent subsets, viz. $\{\{1\}, \{3\}, \{4\}, \{2, 5\}\}$, $\{\{2\}, \{3\}, \{5\}, \{1, 4\}\}$ and $\{\{2\}, \{4\}, \{5\}, \{1, 3\}\}$. So, $f(4) = 3$. This implies that there are three ways of partitioning V so that the vertices of G can be properly colored using precisely four colors. Finally, $f(5) = 1$. Thus,

$$\begin{aligned} \chi_G(\lambda) &= 2\lambda(\lambda - 1)(\lambda - 2) + 3\lambda(\lambda - 1)(\lambda - 2)(\lambda - 3) \\ &\quad + \lambda(\lambda - 1)(\lambda - 2)(\lambda - 3)(\lambda - 4) \\ &= \lambda(\lambda - 1)(\lambda - 2)(\lambda^2 - 4\lambda + 5) \\ &= \lambda^5 - 7\lambda^4 + 19\lambda^3 - 23\lambda^2 + 10\lambda \end{aligned}$$

Hence, the Chromatic number $\chi(G) = 3$, i.e., the given graph G is 3-Chromatic.

Example 8.2 Find the chromatic polynomial of the graph in Fig. 8.4.

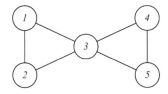

Fig. 8.4

Solution:

Let $f(r)$ be the number of different ways of partitioning the vertex set $V = \{1, 2, 3, 4, 5\}$ into r independent subsets. Since, there is at least one edge in the graph. So, $f(1) = 0$. It is not possible to partition V into two independent subsets. So, $f(2) = 0$. In fact, the vertex set V cannot be partitioned into one or two independent subsets so that the vertices of G can be properly colored using one or two colors, respectively.

There are two ways of partitioning V into three independent subsets, viz. $\{\{1,5\}, \{2,4\}, \{3\}\}$ and $\{\{1,4\}, \{2,5\}, \{3\}\}$. So, $f(3) = 2$. Consequently, there are two ways of partitioning V so that the vertices of G can be properly colored using precisely three colors.

Similarly, there are four ways of partitioning V into four independent subsets, viz. $\{\{2\}, \{3\}, \{4\}, \{1,5\}\}, \{\{1\}, \{3\}, \{5\}, \{2,4\}\}, \{\{1\}, \{3\}, \{4\}, \{2,5\}\}$ and $\{\{2\},\{3\},\{5\},\{1,4\}\}$. So, $f(4) = 4$. This implies that there are four ways of partitioning V so that the vertices of G can be properly colored using precisely four colors. Finally, $f(5) = 1$. Thus,

$$\chi_G(\lambda) = 2\lambda(\lambda - 1)(\lambda - 2) + 4\lambda(\lambda - 1)(\lambda - 2)(\lambda - 3) + \lambda(\lambda - 1)(\lambda - 2)(\lambda - 3)(\lambda - 4)$$
$$= \lambda(\lambda - 1)^2(\lambda - 2)^2$$

Hence, the Chromatic number $\chi(G) = 3$, i.e., the given graph G is 3-Chromatic.

Theorem 8.5 *A graph with n vertices is a complete graph if and only if its Chromatic Polynomial is* $\chi_G(\lambda) = \lambda(\lambda - 1)(\lambda - 2) \cdots (\lambda - n + 1)$.

Proof Using λ colors, there are λ different ways of coloring any selected vertex of the graph. A second vertex can be properly colored in exactly $(\lambda - 1)$ ways, the third vertex in $(\lambda - 2)$ ways, the fourth in $(\lambda - 3)$ ways,... and therefore, the n th vertex in $(\lambda - n + 1)$ ways if and only if every vertex is adjacent to each other. It is possible if and only if the graph G is complete. □

8.4 Edge Contraction

Let G be a graph with $e = uv \in E(G)$ and let $x = x(uv)$ be a new contracted vertex. The graph $G * e$ on

$$V(G * e) = (V(G) - \{u, v\}) \cup \{x\}$$

is obtained from G by contracting the edge e.
    Here,

$$E(G * e) = \{f | f \in E(G), f \text{ has no end } u \text{ or } v\} \cup \{wx | wu \in E(G) \text{ or } wv \in E(G)\}$$

Hence, $G * e$ is obtained by introducing a new vertex x as shown in Fig. 8.5, and by replacing all edges wu and wv by wx, and the vertices u and v are deleted.

Fig. 8.5 An edge contracted graph $G * e$ obtained from G

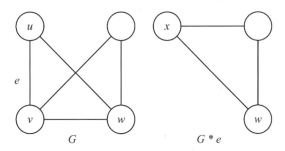

Theorem 8.6 *If T be a tree with n vertices then its Chromatic Polynomial is*

$$\chi_T(\lambda) = \lambda(\lambda - 1)^{n-1}$$

Proof We shall use induction on n. For $n \le 2$, the claim is obvious. Suppose that, $n \ge 3$ and let us assume that the result is true for all trees with $n - 1$ vertices. Let $e = uv \in E(T)$, where v is a leaf of T. A proper λ coloring of T is a proper λ coloring of $T - e$ if and only if the end vertices u and v of e have distinct colors.

    Therefore, we can obtain the proper λ coloring of T by subtracting from proper λ coloring of $T - e$, the number of proper λ coloring of $T - e$ in which u and v have the same color. Now, colorings of $T - e$ in which u and v have the same color correspond directly to proper λ coloring of $T * e$, in which the color of the contracted vertex is the common color of u and v (Fig. 8.6).

Fig. 8.6 **a** A tree T with an
edge e, **b** Disconnected graph
T-e after deleting the edge e,
and **c** Tree $T * e$ after
contracting edge e

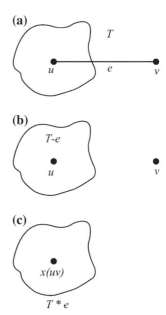

Now, $T * e$ is a tree with $n - 1$ vertices, and thus by induction hypothesis,

$$\chi_{T*e}(\lambda) = \lambda(\lambda - 1)^{n-2}$$

Again, the graph $T - e$ consists of the isolated vertex v and a tree with $n - 1$ vertices. Therefore, since isolated vertex v can be colored in λ different ways

$$\chi_{T-e}(\lambda) = \lambda.\lambda(\lambda - 1)^{n-2}$$

Now,

$$\chi_T(\lambda) = \chi_{T-e}(\lambda) - \chi_{T*e}(\lambda)$$

$$= \lambda.\lambda(\lambda - 1)^{n-2} - \lambda(\lambda - 1)^{n-2}$$

$$= \lambda(\lambda - 1)^{n-1}$$

Hence, it is proved.   □

Critical graphs: A k-chromatic graph G is said to be *k-critical*, if $\chi(H) < k$ for all $H \subseteq G$ with $H \neq G$.

Theorem 8.7 *If G is k-critical for $k \geq 2$, then it is connected, and $\delta(G) \geq k - 1$.*

Proof We can see that for any graph G with the connected components $G_1, G_2, ..., G_m$, $\chi(G) = \max\{\chi(G_i) | i \in [1, m]\}$. Connectivity claim follows from this observation.

Then let G be k-critical, but $\delta(G) = d(v) \le k - 2$ for $v \in G$. Since G is critical, there is a proper $(k - 1)$-coloring of $G - v$. Now v is adjacent to only $\delta(G) < k - 1$ vertices. But there are k colors, and hence there is an available color i for v. If we recolor v by i, then a proper $(k - 1)$-coloring is obtained for G; a contradiction. Hence the theorem is proved. $\square$

Exercises:

1. Find the chromatic polynomial of the graph in Fig. 8.7 and hence find the chromatic number of the given graph.

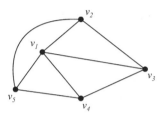

Fig. 8.7

2. Show that if G contains exactly one odd cycle, then $\chi(G) = 3$.
3. If G is a graph in which any pair of odd cycles have a common vertex, then prove that $\chi(G) \le 5$.
4. Determine the chromatic number of the graphs in Fig. 8.8

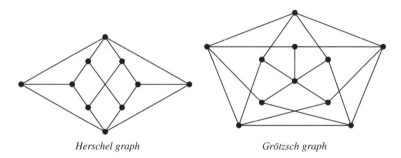

Herschel graph   Grötzsch graph

Fig. 8.8   Herschel and Grötzsch graphs

5. If G is k-regular, prove that $\chi(G) \ge \dfrac{n}{n - k}$.
6. Show that, the chromatic polynomial of a graph consisting of a single circuit of length n is $\chi_G(\lambda) = (\lambda - 1)^n + (\lambda - 1)(-1)^n$.
7. Show that, the chromatic polynomial of a graph of n vertices satisfies the inequality $\chi_G(\lambda) \le \lambda(\lambda - 1)^{n-1}$.
8. Show that, the absolute value of the second coefficient of λ^{n-1} in the chromatic polynomial $\chi_G(\lambda)$ of a graph equals the number of edges in the graph.

Chapter 9
Planar and Dual Graphs

9.1 Plane and Planar Graphs

9.1.1 Plane Graph

Definition A graph is called a plane graph if it can be drawn on a plane in such a way that any two of its edges either meet only at their end vertices or do not meet at all.

9.1.2 Planar Graph

Definition A graph which is isomorphic to a plane graph is called a *planar graph*.

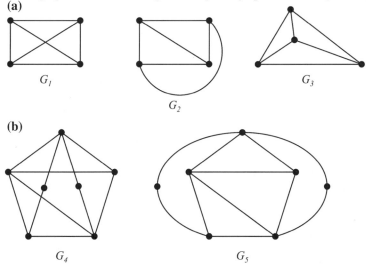

Fig. 9.1 **a** A planar graph G_1 and plane graphs G_2 and G_3 and **b** A planar graph G_4 and plane graph G_5

S. Saha Ray, *Graph Theory with Algorithms and its Applications*,   135
DOI: 10.1007/978-81-322-0750-4_9, © Springer India 2013

Illustration: In Fig. 9.1, G_1 is a planar graph, since it is isomorphic to G_2 and G_3 which are plane graph. Here, G_1 and G_4 are not plane graph. But G_4 is isomorphic to G_5 which is a plane graph. So, G_4 is a planar graph.

9.2 Nonplanar Graph

A graph which is not planar is called nonplanar graph.

Fig. 9.2 A nonplanar graph

The graph shown in Fig. 9.2 cannot be redrawn without crossing over the edges. So, this graph is a nonplanar graph.

Observation

To declare a graph as a planar graph we must see whether the given graph can be redrawn on a plane so that no edges intersect each other. If it is not possible then the graph is nonplanar graph.

It is clear that plane graph is always planar graph.

9.3 Embedding and Region

9.3.1 Embedding

A drawing of a geometric representation of a graph on any surface is called *embedding* such that no edges intersect each other. Therefore, to declare a graph G as a nonplanar graph, we must see, out of all possible geometric representations of the graph G, none can be embedded on a plane.

A graph G is said to be a planar graph if there exists a graph isomorphic to G such that it is embedded on a plane. Otherwise, the graph G is nonplanar.

9.3.2 Plane Representation

An embedding of a planar graph G on a plane is called a *plane representation of G*.

Example In Fig. 9.1, G_5 is a plane representation of G_4.

9.4 Regions or Faces

A plane representation of a graph divides the plane into *regions* or *faces*. A *region* is characterized by the set of edges forming its boundary.

For example: In Fig. 9.3a, there are four regions. On the other hand, Fig. 9.3b has six regions.

Fig. 9.3   a A planar graph K_4 and its corresponding plane graph and **b** A plane graph citing its six regions

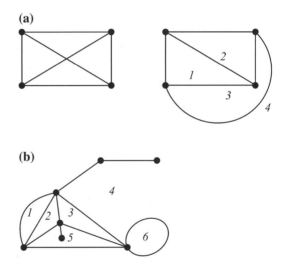

Infinite Region: The portion of the plane lying outside the graph embedded on a plane is called infinite region for that particular plane representation.

For example, in Fig. 9.3a, region 4 is infinite in its extent.

A region is not defined in a nonplanar graph or even in a planar graph not embedded on a plane. For example, the planar graph shown in Fig. 9.3a has no region.

9.5 Kuratowski's Two Graphs

The complete graph K_5 and the complete bipartite graph $K_{3,3}$ are called Kuratowski's graphs, after the polish mathematician **Kazimierz Kuratowski**, who found that K_5 and $K_{3,3}$ are nonplanar.

9.5.1 Kuratowski's First Graph

Fig. 9.4 Kuratowski's first
graph K_5

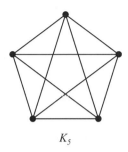

K_5

The complete graph with five vertices, i.e., K_5 is called *Kuratowski's First graph*. In Fig. 9.4, K_5 is nonplanar since plane representation of it is not possible.

9.5.2 Kuratowski's Second Graph

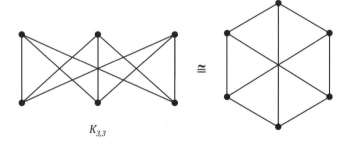

$K_{3,3}$ $\cong$

Fig. 9.5 Kuratowski's second graph $K_{3,3}$

The complete bipartite graph $K_{3,3}$ is called *Kuratowski's second graph*. It is also nonplanar, since plane representation of $K_{3,3}$ is not possible. Figure 9.5 shows Kuratowski's second graph $K_{3,3}$ and its corresponding isomorphic graph.

9.6 Euler's Formula

Theorem 9.1 *A connected planar graph G with n vertices and e number of edges has $f = (e - n + 2)$ number of regions or faces.*

Proof We shall prove this theorem using induction on region f.

If $f = 1$, then G has only one region which is the only infinite region. So, G cannot have any circuit because a circuit bounds a region. So, G is a tree. Now, G has n vertices and $(n - 1)$ edges.

Therefore, $n - e + f = n - (n - 1) + 1 = 2$
Hence, the theorem is proved for $f = 1$.
Now, let $f > 1$ and the theorem is true for all connected planar graph having less than f regions. Since, $f > 1$, G is not a tree. G has at least one circuit.

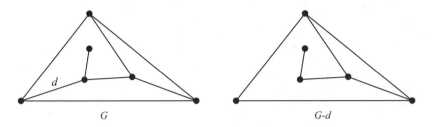

Fig. 9.6 Graph G with four faces and Graph G—d with three faces

Let d be an edge of a circuit. Then $(G - d)$ is still a connected graph. Since $(G - d)$ is a subgraph of planar graph G, so $(G - d)$ is also a planar graph because every subgraph of a planar graph is also a planar graph.

Now, due to removal of an edge d the two regions of the graph G will be combined into one region and so, the number of regions in $(G$—$d)$ is $(f - 1)$ and the number of edges of $(G - d)$ is $(e - 1)$.

Therefore, by induction hypothesis, the theorem is true for $(G - d)$. So, in case of $(G - d)$, we can get

$$n - (e - 1) + f - 1 = 2 \quad \text{i.e.} \quad n - e + f = 2.$$

Hence, by induction, the theorem is true for all connected planar graphs. □

The above Fig. 9.6 shows two faces combines when a cycle edge is deleted in G.

Theorem 9.2 *In any simple planar graph G with f regions, n vertices and e edges where $n \geq 3$ the following inequalities must hold*

(i) $e \geq 3f/2$

(ii) $e \leq 3n - 6$

Proof We first assume that the graph G is connected.
    Case I: If $n = 3$ then the graph G must have at most three edges, i.e., $e \leq 3$ this implies $e \leq 3 \times 3 - 6 = 3n - 6$. Therefore, the result is true for $n = 3$.

If G has exactly 3 edges then G has 2 regions, otherwise G has one region, i.e., if G has $e = 3$ number of edges then $f = 2$, and also if G has $e = 2$ number of edges then $f = 1$ (since, G is simple graph). In either case $e \geq 3f/2$.

Case II:

We now assume that $n \geq 4$.

If G is a tree then G has $e = (n-1)$ edges and $f = 1$, since G is a tree having one region. So, $e = n - 1 \geq 4 - 1 = \frac{3}{2} \times 2 > \frac{3}{2} \times 1 = \frac{3}{2} \times f$. Consequently, $e \geq 3f/2$.

If G is not a tree then there exists at least one circuit whose edges are the boundary of the infinite region. Now, since G is a simple graph so the number of boundary edges of each region of $G \geq 3$.

Therefore the sum of the number of boundary edges of all regions of

$$G \geq 3f \tag{9.1}$$

Consequently, the total number of boundary edges of all regions of $G \leq 2e$. In fact, each edge is counted twice since each edge belongs to exactly two regions. So, from Eq. (9.1) we get

$$3f \leq 2e$$
$$\Rightarrow e \geq \frac{3}{2}f \tag{9.2}$$

By *Euler's formula*, we know that $f = e - n + 2$

Here,

$$e \geq \frac{3}{2}f$$
$$\Rightarrow e \geq 3(e - n + 2)/2$$
$$\Rightarrow 2e \geq 3e - 3n + 6$$
$$\Rightarrow e \leq 3n - 6$$

Now, if possible, suppose G is a disconnected graph. Then G has $G_1, G_2, \ldots, G_k$ connected components.

Let n_i and e_i be the number of vertices and number of edges of the ith component G_i, where $1 \leq i \leq k$, and also, $n = \sum_{i=1}^{k} n_i$, and $e = \sum_{i=1}^{k} e_i$. From the above argument, we can obtain for the ith component of the graph, i.e., G_i,

$$e_i \leq 3n_i - 6 \quad (\text{since, } G_i \text{ is connected})$$
$$\Rightarrow \sum_{i=1}^{k} e_i \leq 3 \sum_{i=1}^{k} n_i - 6k$$
$$\Rightarrow e \leq 3n - 6k \leq 3n - 6$$

Hence, the theorem is proved.   □

Corollary 1 *Prove that K_5 is a nonplanar graph.*

Proof The graph K_5 is a complete graph with five vertices. Here, $n = 5$,

$$e = \frac{n(n-1)}{2}$$
$$= 10$$

From the above theorem 9.2,

$$e \leq 3n - 6$$
$$\Rightarrow e \leq 3 \times 5 - 6 = 9$$

It is a contradiction, since here $e = 10$. So, K_5 is a nonplanar graph.   □

Corollary 2 *Prove that $K_{3,3}$ is a nonplanar graph.*

Proof Here, the number of vertices $n = 6$, and the number of edges $e = 9$.
From the above theorem 9.2,

$$e \leq 3n - 6 = 3 \times 6 - 6 = 12$$

Since, $e = 9$. $e \leq 3n - 6$ implies that $9 < 12$. So, the inequality holds in this case.

But still $K_{3,3}$ is a nonplanar graph.

We know that $K_{3,3}$ is a complete bipartite graph and so it has no odd cycle. In particular, it has no 3-cycle. It follows that every region of a plane drawing of $K_{3,3}$, if such exists, must have at least four boundary edges. Therefore the number of boundary edges of each region of $K_{3,3} \geq 4$.

Consequently, the sum of number of boundary edges of all regions of

$$K_{3,3} \geq 4f. \tag{9.3}$$

Now, the sum of number of boundary edges of all regions of $K_{3,3} \leq 2e$, since, each edge is counted twice.

$$\text{Therefore, } 4f \leq 2e \quad \text{i.e.} \quad f \leq e/2 = 9/2. \tag{9.4}$$

Now, Suppose $K_{3,3}$ is a planar graph; therefore, the plane representation of $K_{3,3}$ is possible.

By Euler's formula, we know $f = e - n + 2$. This implies $f = 9 - 6 + 2 = 5$ which is a contradiction in view of Eq. (9.4).

Hence, $K_{3,3}$ is a nonplanar graph.   □

Example 9.1 Removal of one edge or vertex makes Kuratowski's first graph a planar graph.

Solution:

Removal of one edge

In Fig. 9.7a, G represents the Kuratowski's first graph. $G - e$ is the graph obtained after removal of edge e from G shown in Fig. 9.7b. Finally, Fig. 9.7c shows the plane representation of the graph $G - e$. So, $G - e$ is a planar graph.

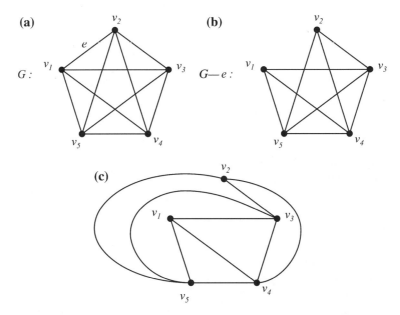

Fig. 9.7 **a** Nonplanar graph G, **b** Planar graph $G–e$, and **c** Plane representation of $G–e$

Removal of one vertex

Removal of vertex v_1 from G in Fig. 9.7a results in a graph $G - v_1$ shown in Fig. 9.8a. $G - v_1$ is redrawn below without any crossover of edges. Figure 9.8b shows the plane representation of $G - v_1$. This shows that $G - v_1$ is a planar graph.

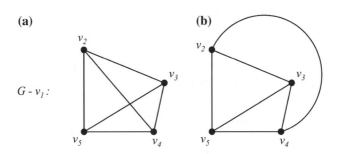

Fig. 9.8 **a** Planar graph $G - v_1$ and **b** Plane representation of $G - v_1$

9.7 Edge Contractions

Let G be a multigraph, and let e be an edge of G with distinct endpoints x and y.
The Contraction $G * e$ or $G|e$ can be defined by modifying G as follows:
We remove the edge e and identify its endpoints x and y to obtain one vertex v_e
as shown in Fig. 9.9.

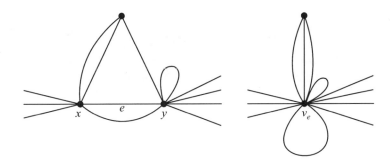

Fig. 9.9 Graphs showing edge contraction

If $e' = ux$ is an edge in G with x as an endpoint, it is replaced by the edge uv_e in
$G|e$, and likewise for edges with y as an endpoint. Loops at one of x or y become
loops at v_e. This may also introduce new loops or parallel edges.

In fact, if G is planar, then $G|e$ is also planar.

9.8 Subdivision, Branch Vertex, and Topological Minors

Let G, H be multigraphs, and $e \in E(G)$.

Subdivision of edge: The *subdivision of edge* $e = xy$ is the replacement of e with
a new vertex z and the two new edges xz and zy.

Subdivision: A (multi-)graph H' is a *subdivision* of H, if one can obtain H' from
H by a series of edge subdivisions (Fig. 9.10)

Fig. 9.10 A subdivision of
$K_{3,3}$

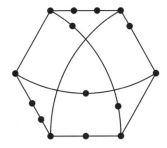

Branch Vertices: Vertices of H' with degree at least three are called branch vertices.

Minor: If H can be obtained from G by a sequence of contractions and (edge/vertex) deletions, then H is called a minor of G.

Topological Minors: If G contains a subdivision of H, then H is called a topological minor of G.

A topological minor is a minor, but not vice versa.

Characterization of Planar Graphs:

The Polish Mathematician **Kazimierz Kuratowski** discovered an interesting property of planar and nonplanar graphs. In his honor, the two graphs K_5 and $K_{3,3}$ are called Kuratowski's first graph and Kuratowski's second graph, respectively.

It has been proved that the complete graph K_5 and the complete bipartite graph $K_{3,3}$ are nonplanar. K_5 and $K_{3,3}$ do not embed in the plane. In fact, these are the crucial graphs and lead to a characterization of planar graphs known as Kuratowski's theorem.

The following result is used in proving Kuratowski's theorem.

Theorem 9.3

i. *If $G|e$ contains a subdivision of K_5, then G contains a subdivision of K_5 or $K_{3,3}$.*

ii. *If $G|e$ contains a subdivision of $K_{3,3}$, then G contains a subdivision of $K_{3,3}$.*

Proof Let $G' = G|e$ be a graph obtained by contracting the edge $e = xy$ of G. Let w be the vertex of G' obtained by contracting $e = xy$.

i. Let $G|e$ contains a subdivision of K_5, say H. If w is not a branch vertex of H, then G also contains a subdivision of K_5, obtained by expanding w back into the edge xy, if necessary, as shown in Fig. 9.11.

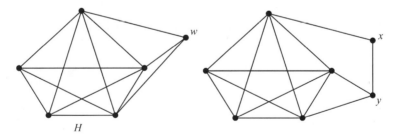

Fig. 9.11

Assume w is a branch vertex of H and each of x, y is incident in G to two of the four edges incident to w in H. Let u_1 and u_2 be the branch vertices of H that are at the other ends of the paths leaving w on edges incident to x in G. Let v_1, v_2 be the branch vertices of H that are at the other ends of the paths leaving w on edges incident to y in G (Fig. 9.12a).

(a)

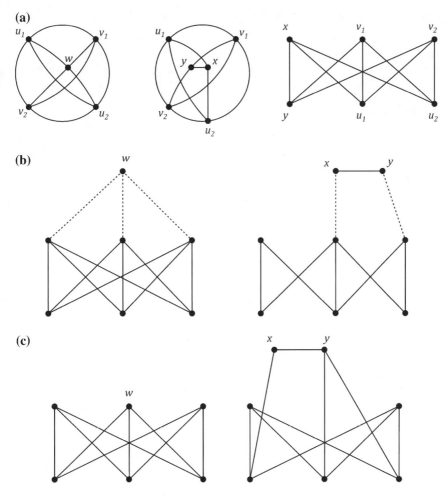

(b)

(c)

Fig. 9.12

By deleting the $u_1 - u_2$ path and $v_1 - v_2$ path from H, we obtain a subdivision of $K_{3,3}$ in G, in which y, u_1, u_2 are branch vertices for one partite set, and x, v_1, v_2 are branch vertices of the other.

ii. Let $G|e$ contains a subdivision of $K_{3,3}$, say H. If w is not a branch vertex of H, then G also contains a subdivision of $K_{3,3}$, obtained by expanding w back into the edge xy, if necessary (Fig. 9.12b).

Now, assume that w is a branch vertex in H and at most one of the edges incident to w in H is incident to x in G. Then w can be expanded into xy to lengthen that path and y becomes the corresponding branch vertex of $K_{3,3}$ in G (Fig. 9.12c). □

9.9 Kuratowi's Theorem

This theorem was independently given by Kuratowski in 1930. In 1954, Dirac and Schuster found a poof that was slightly shorter than the original proof. The proof given here is due to Thomassen (1981).

Theorem 9.4 *A graph is planar if and only if it does not have any subdivision of K_5 or $K_{3,3}$.*

Proof Necessity: Let G be a planar graph. Then any of its subgraphs is neither K_5 nor $K_{3,3}$ nor does it contain any subdivision of K_5 or $K_{3,3}$.

Sufficiency: It is enough to prove sufficiency for 3-connected graphs. Let G be a 3-connected graph with n vertices. We prove that the 3-connected graph G either contains a subdivision of K_5 or $K_{3,3}$ or has a convex plane representation. This we prove by using induction on n. Since G is 3-connected, therefore, $n \geq 4$. For $n = 4$, $G = K_4$ and clearly G has a plane representation.

Now, let $n \geq 5$. Assume the result to be true for all 3-connected graphs with fewer than n vertices. Since G is 3-connected, G has an edge e such that $G|e$ is 3-connected. Let $e = xy$.

If $G|e$ contains a subdivision of K_5 or $K_{3,3}$, then G also contains a subdivision of K_5 or $K_{3,3}$.

Therefore, let $G|e = H$ have a convex plane representation. Let z be the vertex obtained by contraction of $e = xy$. The plane graph obtained by deleting the edges incident to z has a region containing z (this may be the exterior region). Let C be the cycle of $H - z$ bounding this region.

Since we started with a convex plane representation of H, we have straight segments from z to all its neighbors. Let $x_1, x_2, ..., x_k$ be the neighbors of x in that order on C.

If all the neighbors of y belong to a single segment from x_i to x_{i+1} on C, then we obtain a convex plane representation of G by putting x at z in H, and putting y at a point close to z in the wedge formed by xx_i and xx_{i+1}.

If all the neighbors of y do not belong to any single segment $x_i\, x_{i+1}$ on C $(1 \leq i \leq k, x_{k+1} = x_1)$, then we have the following cases as shown in Fig. 9.13.

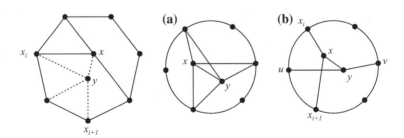

Fig. 9.13

1. y shares three neighbors with x. In this case C together with these six edges involving x and y form a subdivision of K_5.
2. y has two u, v in C that are in different components of the subgraph of C obtained by deleting x_i and x_{i+1}, for some i. In this case, C together with the paths uyv, $x_i x x_{i+1}$, and xy form a subdivision of $K_{3,3}$.   □

Homeomorphic Graphs: Two graphs are said to be homeomorphic if one graph can be obtained from the other by the creation of edges in series or by the merging of edges in series.

Equivalently, two graphs are homeomorphic if they are both subdivisions of some graph.

For Example:

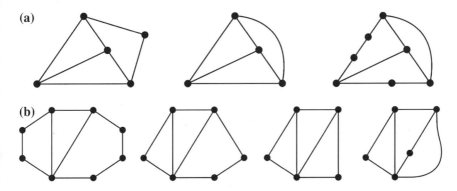

Fig. 9.14 Homeomorphic graphs

The three graphs in Fig. 9.14a are homeomorphic to each other. Similarly, all the graphs in Fig. 9.14b are homeomorphic to each other.

Applications of *Kuratowski's Theorem*:

Formally, we can state Kuratowski's theorem as:

A graph G is planar if and only if G does not contain either of the Kuratowski's two graphs or any graph homeomorphic to either of them.

Remarks   If the contracted graph $G|e$ contains a subdivision of $K_{3,3}$ then so does G. If $G|e$ contains a subdivision of K_5 then G contains a subdivision of K_5 or $K_{3,3}$ (It does not need to contain a subdivision of K_5).

A further result is that:

A graph is planar if and only if it contains no subgraph contractible to K_5 or $K_{3,3}$ by removing edges from the subgraph and merging the adjacent vertices into one.

For example, we can contract the Petersen graph as shown below, thus proving the Petersen graph to be nonplanar (Fig. 9.15).

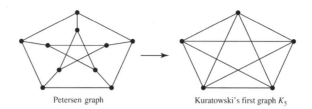

Fig. 9.15 Petersen graph which is Homeomorphic to K_5

Example 9.1 Using Kuratowski's theorem, check if the following graphs in Fig. 9.16 are nonplanar.

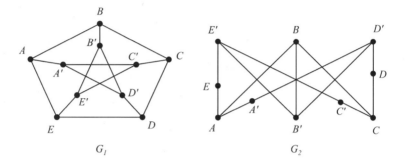

Fig. 9.16 Petersen graph G_1 and its subgraph G_2 which is Homeomorphic to $K_{3,3}$

Solution:

The graph G_1 has a subgraph G_2 and clearly G_2 is Homeomorphic to $K_{3,3}$.

Therefore, according to Kuratowski's theorem, the given graphs G_1 and G_2 are nonplanar graphs.

Example 9.2 Using Kuratowski's theorem, check that the following graphs are nonplanar (Fig. 9.17).

Fig. 9.17 Two nonplanar graphs G_1 and G_2

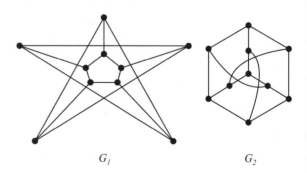

Solution:

The graph G_1 is isomorphic to G_2 and it can be shown that G_2 is Homeomorphic to $K_{3,3}$. So according to Kuratowski's theorem, the given graphs both G_1 and G_2 are nonplanar graphs.

Example 9.3 If K_n is a complete graph with n vertices then find all integral values of $n \geq 2$ for which K_n is a planar graph.

Solution:

If $n = 2, 3$ then K_2, K_3 are planar graphs respectively.

Similarly,

If $n = 4$, then K_4 is a planar graph as shown in Fig. 9.3a.

Now, if $n = 5$, then we have proved that K_5 is a nonplanar graph.

If $n = 6$, then according to Kuratowski's theorem, the graph K_6 is a nonplanar graph since K_6 contains K_5 as a subgraph (Fig. 9.18).

Fig. 9.18 Nonplanar complete graph K_6

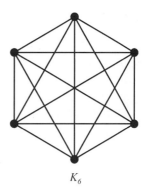

K_6

Similarly, according to Kuratowski's theorem, K_7 is nonplanar since K_7 also contains K_5 as a subgraph, and so on.

Hence, the required integral values of n are 2, 3, and 4.

9.10   Dual of a Planar Graph

The following Fig. 9.19a shows the plane representation of a graph G with six regions or faces F_1, F_2, F_3, F_4, F_5, and F_6.

9.10.1   To Find the Dual of the Given Graph

First: Let us place six points $P_1, P_2, \ldots, P_6$ one in each of the regions, as shown in Fig. 9.19b.

Fig. 9.19 Graph G and its Dual graph G^*

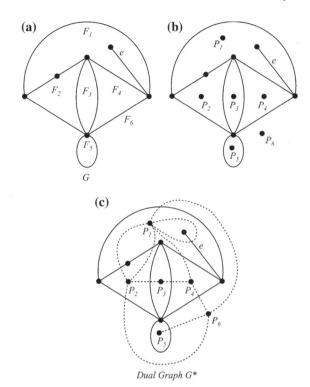

Second: Let us join these 6 points according to the following procedure.

Construction of dual graph: If two regions F_i and F_j are adjacent (i.e., have a common edge), draw a line joining the points P_i and P_j such that the line joining these points intersect the common edge between the regions F_i and F_j exactly once. For example, in Fig. 9.19c, F_3 and F_4 are adjacent. So, we join P_3 and P_4 by an edge.

If there is more than one edge common between F_i and F_j, draw one line between points P_i and P_j for each of the common edges. For example, in Fig. 9.19c, there are two edges common between F_1 and F_2. So, we draw one edge between P_1 and P_2 for each of the common edges.

If an edge e lies entirely in one region say F_k, a self-loop is to be drawn at the point P_k intersecting the edge e exactly once. So, we draw a loop at the point P_1 intersecting the edge e exactly once.

By this procedure, we obtain a new graph G^* from the given graph G consisting of six vertices $P_1, P_2, \ldots, P_6$ corresponding to the regions $F_1, F_2, \ldots, F_6$ of G and the edges joining these six vertices $P_1, P_2, \ldots, P_6$. Such a graph G^* is called a dual of the graph G.

9.10.2 Relationship Between a Graph and Its Dual Graph

1. There is a one-to-one correspondence between the edges of the graph G and its dual G^*, since one edge of G^* intersects one edge of G exactly once.
2. Dual graph is possible only for planar graph. Dual of G^* is the original graph G.
3. An edge forming a self-loop in G yields a pendant edge in G^*. A pendant edge is an edge incident on a pendant vertex. A pendant edge in G yields a self-loop in G^*.
4. Edges that are in series in G produce parallel edges in G^*. Parallel edges in G produce edges in series in G^*.
5. The number of vertices of G^* is equal to number of regions of G.
6. The number of edges of G^* is equal to number of edges of G, since one edge of G^* intersects one edge of G exactly once.
7. The number of regions of G^* is equal to number of vertices of G.

Example 9.4 Show that the given two plane graphs G_1 and G_2 are isomorphic but their dual are not isomorphic.

Fig. 9.20 Two isomorphic graphs G_1 and G_2

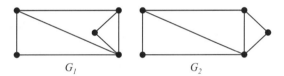

G_1 G_2

Solution:
From Fig. 9.20, we see that the number of edges and the number of vertices are the same in G_1 and G_2. Also, incidence property is preserved.

Therefore, $G_1 \cong G_2$

Dual of G_1 and G_2:

The dual graphs G_1^* and G_2^* are not isomorphic because the incidence property is not preserved (Fig. 9.21).

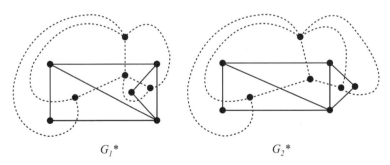

$G_1{}^*$ $G_2{}^*$

Fig. 9.21 Dual graphs $G_1{}^*$ and $G_2{}^*$

Example 9.5 Using Kuratowski's theorem, show that the following graphs below are nonplanar (Fig. 9.22).

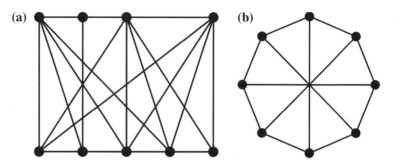

Fig. 9.22   Two nonplanar graphs

Solution:
The graph in Fig. 9.23a has a subgraph shown in Fig. 9.23b which is $K_{3,3}$. Therefore, by Kuratowski's theorem, the graph in Fig. 9.23a is nonplanar.

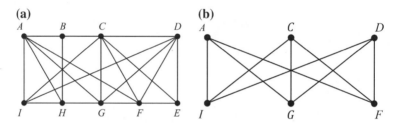

Fig. 9.23

Figure 9.24a has a subgraph shown in Fig. 9.24b. Now, Fig. 9.24b is homeomorphic to Fig. 9.24c. Again, Fig. 9.24c is isomorphic to Fig. 9.24d which is $K_{3,3}$. Therefore, by Kuratowski's theorem, the graph in Fig. 9.24a is nonplanar.

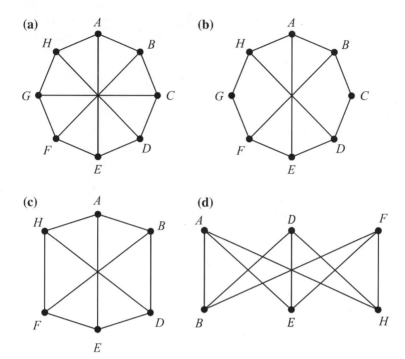

Fig. 9.24

9.11 Edge Coloring

Assignment of colors to the edge of a graph G, so that no two adjacent edges receiving the same color is called an Edges coloring of G. k-edge coloring of a graph G is an assignment of k-colors to the edge of G such that no two edges of G receive the same color.

9.11.1 k-Edge Colorable

A graph G is said to be k-edge colorable, if there exists k-edge coloring of G.

9.11.2 Edge-Chromatic Number

The minimum number k, such that a graph G has k-edge coloring is said to be the edge-chromatic number of G. The edge-chromatic number of a graph G is denoted by $\chi'(G)$. Thus, $\chi'(G)$ denotes the minimum number of colors required to color the edges of the graph G, such that no two adjacent edges of G receive the same color.

9.12 Coloring Planar Graph

The most famous problem in the history of graph theory is that of the chromatic number of planar graphs. The problem was known as the 4-*Color Conjecture* for more than 120 years, until it was solved by Appel and Haken in 1976; if G is a planar graph, then $\chi(G) \leq 4$. The 4-color Conjecture has had a deep influence on the theory of graphs during the last 150 years. The solution of the 4-color Theorem is difficult, and it requires the assistance of a computer.

Four-Color Problem:

The 4-color problem states that every plane map however complex, can be colored with four colors in such a way that two adjacent regions get different colors. This problem was solved by Appel and Haken in 1976.

Four-Color Conjecture:

Every planar graph is 4-colorable.

Illustration:

The graph K_4 is a planar graph and K_4 is 4-colorable as shown in Fig. 9.25.

Fig. 9.25 A planar graph with its four regions colored

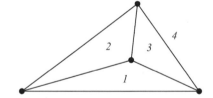

9.12.1 The Four Color Theorem

The Four Color Theorem states that given any separation of a two-dimensional area into connected regions, called a map, the regions can be colored using at most four colors so that no two adjacent regions have the same color.

Note: It is important to realize that two regions are called adjacent only if they share a border segment, and not just a point.

The now famous theorem was first conjectured in 1852, but was not proven until 1976, by Appel and Haken. They determined that there are 1936 ways to draw a map (all others being equivalent to one of them), and that after thousands of hours of computation they had reached the conclusion that in each case only four colors are needed. Needless to say, many were skeptical of this method of proof, but after the 400 pages of microfiche output were independently checked, it was declared as being valid. Thus the proof, unlike most proofs in mathematics, is technology dependent; that is, it uses computers in an essential way and depended on the development of high-speed computers.

9.12.2 The Five Color Theorem

We prove Heawood's result (1890) that each planar graph is properly 5-colorable. The proof of the following theorem 9.6 is partly geometric in nature.

Theorem 9.5 (Heawood 1890) *If G is a planar graph, then* $\delta(G) \leq 5$.

Proof If $n \leq 2$, then there is nothing to prove. Suppose $n \geq 3$. By the Handshaking lemma and theorem 9.2

$$\delta(G).n \leq \sum_{v \in G} d(v) = 2e \leq 6n - 12 < 6n$$

It follows that $\delta(G) \leq 5$. ☐

Theorem 9.6 (Heawood (1890)) (*Five Color Theorem*) *Every planar graph is 5-colorable, i.e., if G is a planar graph, then* $\chi(G) \leq 5$.

Proof It is equivalent to prove that no 6-critical planar graph exists. From theorems 8.7 and 9.5, it follows that 6-critical graph G must have $\delta(G) \geq 5$ and planar graph must have $\delta(G) \leq 5$. We can then assume G has a degree 5 vertex with neighbors using all 5 colors in consecutive order. Denoting these neighbors v_1, v_2, v_3, v_4, v_5. Then we may switch color 1 to 3 on v_1 and make any necessary corrections. If v_3 needs correction, then there must be a 1, 3 path connecting them; but then we can change color 2 to 4 on v_2 and never gets to v_4. Therefore, G cannot be 6-critical. ☐

The final word on the chromatic number of planar graph was proved by Appel and Haken in 1976.

Theorem 9.7 (4-Color Theorem) *Every planar graph is 4-colorable*, i.e., *if G is a planar graph, then* $\chi(G) \leq 4$.

By the following theorem, each planar graph can be decomposed into two bipartite graphs.

Theorem 9.8 *Let* $G = (V, E)$ *be a 4-chromatic graph,* $\chi(G) \leq 4$. *Then the edges of G can be partitioned into two subsets* E_1 *and* E_2 *such that* (V, E_1) *and* (V, E_2) *are both bipartite.*

Proof Let $V_i = \alpha^{-1}(i)$ be the set of vertices colored by i in a proper 4-coloring α of G. We define E_1 as the subset of the edges of G that are between the sets V_1 and V_2; V_1 and V_4; V_3 and V_4. Let E_2 be the rest of the edges, that is, they are between the sets V_1 and V_3; V_2 and V_3; V_2 and V_4. It is clear that (V, E_1) and (V, E_2) are bipartite, since the sets V_i are stable. ☐

9.13 Map Coloring

The 4-color conjecture was originally stated for maps. In the map-coloring problem we are given several countries with common borders, and we wish to color each country so that no neighboring countries obtain the same color.

A border between two countries is assumed to have a positive length—in particular, countries that have only one point in common are not allowed in the map coloring.

Formally, we define a map as a connected planar (embedding of a) graph with no bridges [a bridge (also known as a cut-edge) is an edge whose deletion increases the number of connected components]. The edges of this graph represent the boundaries between countries. Hence a country is face of the map, and two neighboring countries share a common edge (not just a single vertex). We deny bridges, because a bridge in such a map would be a boundary inside a country.

The map-coloring problem is restated as follows:

How many colors are needed for the faces of a plane embedding so that no adjacent faces obtain the same color.

Let $F_1, F_2, \ldots, F_n$ be the countries of a map M, and define a graph G with $V_G = \{v_1, v_2, \ldots, v_n\}$ such that $v_i v_j \in E_G$ if and only if the countries F_i and F_j are neighbors. It is easy to see that G is a planar graph. Using this notion of a dual graph, we can state the map-coloring problem in a new form: *What is the chromatic number of a planar graph?* By the 4-Color theorem it is at most four.

Example: If we look at the map of the United States below, we see that only four colors are used to color the states (Fig. 9.26).

Fig. 9.26 Map of *United States*

Note: We can use only three colors for many maps, a fourth being needed when a region has common borders with an odd number of neighboring regions. A basic example is given below: (Fig. 9.27).

Fig. 9.27 A map with its regions colored with four colors

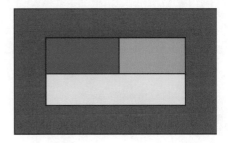

Exercises:

1. Show that all circuit graphs are homeomorphic to C_3.
2. Show that K_3 is homeomorphic to $K_{2,2}$.
3. Suppose G_1 has v_1 vertices and e_1 edges and that G_2 has v_2 vertices and e_2 edges and that G_1 is homeomorphic to G_2. Show that $e_1 - v_1 = e_2 - v_2$.
4. If G is Eulerian and H is homeomorphic to G, is H Eulerian?
5. If G is Hamiltonian and H is homeomorphic to G, is H Hamiltonian?
6. Use Kuratowski's theorem to show that K_n is nonplanar for $n \geq 5$.
7. Using Kuratowski's theorem, show that the following graphs are nonplanar (Fig. 9.28).

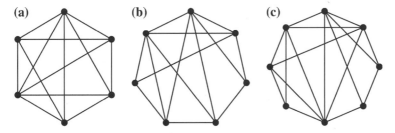

Fig. 9.28

8. Find the dual of the following two graphs (Fig. 9.29).

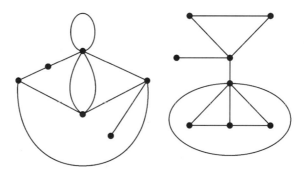

Fig. 9.29

9. Give an example of a graph G, the dual of whose dual is again G. [Hint: complete graph K_4 is self-dual].

10. If G is a planar graph with n vertices, e edges, f regions, and k components then prove that $n - e + f = k + 1$.

11. Prove that if G is self-dual (i.e. G and G^* are same or isomorphic) with n vertices and e edges then $e = 2n - 2$.

12. Let G be a connected simple planar graph. Prove that if $d(v_i) \geq 5$ for all vertex v_i of G, then there are at least 12 vertices of degree 5 in G.

13. If every region of a simple planar graph with n vertices and e edges embedded in a plane is bounded by k edges, show that $e = \frac{k(n-2)}{(k-2)}$.

14. Find a simple graph G with degree sequence [4,4,3,3,3,3] such that
    (a) G is planar.
    (b) G is nonplanar.

15. Show that a set of fundamental circuits in a planar graph G corresponds to a set of fundamental cut sets in its dual G^*.

Chapter 10
Network Flows

Let us consider a network of pipelines of oil, gas, water, and so on. If we consider the case of network of pipes having values allowing flows only in one direction. It is important to note that each pipe has capacity. This type of network is represented by weighted connected graph in which stations are represented by vertices or nodes and lines through which given item such as oil, gas, water, electricity, etc., flows through by edges and capacities by weights. We also assume that flow cannot accumulate at an intermediate level. It is assumed that at each intermediate vertex, the total rate of commodity entering (in-flow) is equal to the rate of leaving (out-flow). One most important thing that will arise in many applications, what is the maximal (or maximum) flow from source vertex (source station) to sink vertex (destination station) in all these types of transmission network.

10.1 Transport Networks and Cuts

10.1.1 Transport Network

A simple, connected, weighted, digraph (directed graph) G is called a Transport Network (Flow Network) if the weighted associated with every directed edge in G is a non-negative number. In a Transport Network, this non-negative number represents the capacity of the edge and it is denoted by c_{ij} for the directed edge (i, j) in G.

To illustrate, Fig. 10.1 shows a Transport network with source vertex s and sink vertex t. The flows are shown in Fig. 10.2. In Fig. 10.2, the net flow out of source s is $10 + 3 + 11 = 24$ units.

In the *maximum-flow problem*, we are given a Transport Network or Flow Network G with source s and sink t, and we wish to find a flow of maximum value from s to t.

S. Saha Ray, *Graph Theory with Algorithms and its Applications*,
DOI: 10.1007/978-81-322-0750-4_10, © Springer India 2013

Fig. 10.1 A transport
network or flow network

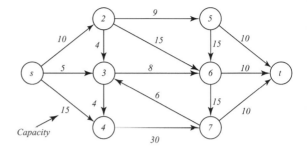

Fig. 10.2 A transport
network showing flows and
corresponding capacities of
edges

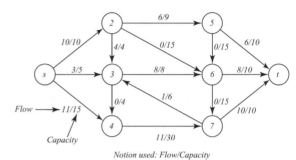

Notion used: Flow/Capacity

Flow:

In a given Transport Network G, a flow is an assignment of a non-negative number f_{ij} to every directed edge (i, j) such that the following conditions are satisfied:

1. For every directed edge (i, j) in G

$$f_{ij} \leq c_{ij} \tag{10.1}$$

This condition implies that flow through any edge does not exceed its capacity.

2. For the source vertex s in G

$$\sum_{i \neq s} f_{si} - \sum_{i \neq s} f_{is} = w \tag{10.2}$$

where the above summations are taken over all vertices in G and w is called value of the flow.

This condition states that the net flow out of the source vertex s is w.

3. For the sink vertex t in G

$$\sum_{i \neq t} f_{ti} - \sum_{i \neq t} f_{it} = -w \tag{10.3}$$

This condition states that the net flow into the sink vertex t is w.

4. All other vertices, apart from source vertex s and sink vertex t, are called intermediate vertices. For each such intermediate vertex j

$$\sum_{i \neq j} f_{ij} = \sum_{i \neq j} f_{ji} \tag{10.4}$$

The above condition implies that the flow is conserved at each intermediate vertex.
Saturated edge:
An edge (i, j) is said to be *saturated* if the flow f_{ij} in edge (i, j) is equal to its capacity c_{ij}, i.e. $f_{ij} = c_{ij}$.
Flow Pattern:
A set of flows f_{ij} for all edge (i, j) in G is called a *Flow Pattern*.
Maximal Flow Pattern:
A flow pattern that maximizes the quantity w, which is the value of the flow from source vertex s to sink vertex t, is called a *Maximal Flow Pattern*.

10.1.2 Cut

A cut is a partition of vertices in G into two non-empty subsets P and $\overline{P}$, where P always contains s and $\overline{P}$ always contains t. It is usually denoted by $(P, \overline{P})$.
Capacity of a Cut:
The capacity of a cut $(P, \overline{P})$, denoted by $c(P, \overline{P})$, is defined as the sum of capacities of those edges directed from the vertices in set P to the vertices in set $\overline{P}$. It is given by

$$c(P, \overline{P}) = \sum_{\substack{i \in P \\ j \in \overline{P}}} c_{ij} \tag{10.5}$$

Theorem 10.1 *In a given Transport Network G, the value of the flow w from source vertex s to sink vertex t is less than or equal to the capacity of any cut in G that separates source vertex s from sink vertex t.*

Proof Let, $(P, \overline{P})$ be any arbitrary cut in G such that the source vertex $s \in P$ and the sink vertex $t \in \overline{P}$.
From Eq. (10.4), for all intermediate vertices $j \in P$

$$\sum_{i \neq j} f_{ji} - \sum_{i \neq j} f_{ij} = 0 \tag{10.6}$$

And from Eq.(10.2), we have

$$\sum_{i \neq s} f_{si} - \sum_{i \neq s} f_{is} = w \tag{10.7}$$

Adding Eqs. (10.6) and (10.7), we get

$$\sum_{\substack{k \in P \\ i \in G}} f_{ki} - \sum_{\substack{k \in P \\ i \in G}} f_{ik} = w \qquad (10.8)$$

The Eq. (10.8) can be rewritten as

$$\sum_{\substack{k \in P \\ i \in P}} f_{ki} + \sum_{\substack{k \in P \\ i \in \overline{P}}} f_{ki} - \sum_{\substack{k \in P \\ i \in P}} f_{ik} - \sum_{\substack{k \in P \\ i \in \overline{P}}} f_{ik} = w \qquad (10.9)$$

Since, the flow is conserved at each intermediate vertex.

$$\sum_{\substack{k \in P \\ i \in P}} f_{ki} - \sum_{\substack{k \in P \\ i \in P}} f_{ik} = 0, \qquad (10.10)$$

Therefore,

$$\sum_{\substack{k \in P \\ i \in \overline{P}}} f_{ki} - \sum_{\substack{k \in P \\ i \in \overline{P}}} f_{ik} = w \qquad (10.11)$$

Since, $\sum_{\substack{k \in P \\ i \in \overline{P}}} f_{ik}$ is always a non-negative quantity.

We obtain $w \leq \sum_{\substack{k \in P \\ i \in \overline{P}}} f_{ki} \leq \sum_{\substack{k \in P \\ i \in \overline{P}}} c_{ki} = c(P, \overline{P})$. Hence it is proved. $\square$

10.2 Max-Flow Min-Cut Theorem

Theorem 10.2 *In a given Transport Network G, the maximum value of the flow, from source vertex s and sink vertex t, is equal to the minimum value of the capacities of all cuts in G that separate source vertex s from sink vertex t. It is given by*

$$\text{Max } w = \min \{ \, c(P, \overline{P}) | (P, \overline{P}) \text{ is any cut in } G \}$$

Proof Using Theorem 10.1, we have

$$w \leq \sum_{\substack{k \in P \\ i \in \overline{P}}} f_{ki} \leq \sum_{\substack{k \in P \\ i \in \overline{P}}} c_{ki} = c(P, \overline{P}) \qquad (10.12)$$

We need only to prove that there exists a flow pattern in G from source vertex s to sink vertex t such that the value of the flow w^*, from source vertex s to sink vertex t, is equal to $c(P^*, \overline{P}^*)$, which is the capacity of some cut $c(P^*, \overline{P}^*)$ separating source vertex s from sink vertex t.

Let, there be some flow pattern in G such that the value of the flow, from source vertex s to sink vertex t, attains its maximum possible value w^*.

We define a vertex set P^* in G recursively as follows:

1. $s \in P^*$
2. If vertex $i \in P^*$ and either $f_{ji} < c_{ij}$ or $f_{ji} > 0$, then $j \in P^*$. If any vertex not in P^*, then it belongs to $\overline{P}^*$.

Now, vertex t is always contained in $\overline{P}^*$. If it can not be contained in $\overline{P}^*$, then there would be a path P_1 from s to t, say $s, v_1, v_2, \ldots, v_j, v_{j+1}, \ldots, v_r, t$ for which in every edge either flow $f_{v_j v_{j+1}} < c_{v_j v_{j+1}}$ or $f_{v_{j+1} v_j} > 0$.

In path P_1 an edge (v_j, v_{j+1}) directed from v_j to v_{j+1} is interpreted as a *forward edge* and an edge (v_{j+1}, v_j) directed from v_{j+1} to v_j is interpreted as a *backward edge*.

In path P_1, let β_1 be the minimum of all differences $c_{v_j v_{j+1}} - f_{v_j v_{j+1}}$ in forward edges and let, β_2 be the minimum of all flows in backward edges. Let $\beta = \min(\beta_1, \beta_2)$, since β_1 and β_2 are positive quantities. Then the flow in the Network G can be increased by increasing the flow in each forward edge and decreasing the flow in each backward edge by an amount β. This contradicts the assumption that w^* was the maximum flow.

Thus t must be in the vertex set $\overline{P}^*$. In other words, the cut $(P^*, \overline{P}^*)$ separates s from t. Again, according to condition (2) for each vertex k in P^* and i in $\overline{P}^*$, we have $f_{ki} = c_{ki}$ and $f_{ik} = 0$.

Therefore, from Eq. (10.11), we get the value of the flow as

$$w^* = \sum_{\substack{k \in P^* \\ i \in \overline{P}^*}} f_{ki} - \sum_{\substack{k \in P^* \\ i \in \overline{P}^*}} f_{ik}$$

$$= \sum_{\substack{k \in P^* \\ i \in \overline{P}^*}} c_{ki} = c(P^*, \overline{P}^*)$$

Hence, the theorem is proved.   □

Example 10.1   Consider the Transport Network in Fig. 10.3, determine the Maximal Flow from source vertex s to sink vertex t.

Fig. 10.3 A transport
network

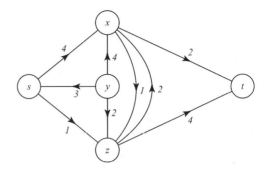

Solution:

There are 3 intermediate vertices between source vertex s and sink vertex t. So, the given Transport Network has $2^3 = 8$ Cuts that separate source vertex s from sink vertex t, since there are two ways for each intermediate vertex to be either included in P or in $\overline{P}$.

Sl. no.	Cut $(P, \overline{P})$	Capacity $c(P, \overline{P})$
1.	$P = \{s\}$, $\overline{P} = \{x,y,z,t\}$	$4 + 1 = 5$
2.	$P = \{s,x\}$, $\overline{P} = \{y,z,t\}$	$1 + 2 + 1 = 4$
3.	$P = \{s,y\}$, $\overline{P} = \{x,z,t\}$	$4 + 4 + 2 \ + \ 1 = 11$
4.	$P = \{s,z\}$, $\overline{P} = \{x,y,t\}$	$4 + 2 + 4 = 10$
5.	$P = \{s,x,y\}$, $\overline{P} = \{z,t\}$	$1 + 1 + 2 + 2 = 6$
6.	$P = \{s,x,z\}$, $\overline{P} = \{y,t\}$	$2 + 4 = 6$
7.	$P = \{s,y,z\}$, $\overline{P} = \{x,t\}$	$4 + 4 + 2 + 4 = 14$
8.	$P = \{s,x,y,z\}$, $\overline{P} = \{t\}$	$2 + 4 = 6$

The minimum capacity among these Cuts occurs for the Cut $(P, \overline{P})$ where $P = \{s,x\}$, $\overline{P} = \{y,z,t\}$. Therefore, according to *Max-Flow Min-Cut theorem*, the maximal flow, from source vertex s to sink vertex t, is 4 units.

Example 10.2 Consider the Transport Network in Fig. 10.4, determine the Maximal Flow from source vertex s to sink vertex t.

Fig. 10.4 A transport
network

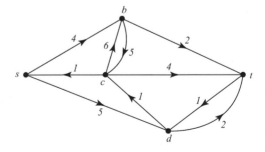

Solution:

There are 3 intermediate vertices between source vertex s and sink vertex t. So, the given Transport Network has $2^3 = 8$ Cuts that separate source vertex s from sink vertex t, since there are two ways for each intermediate vertex to be either included in P or in $\overline{P}$.

Sl. no.	Cut $(P, \overline{P})$	Capacity $c(P, \overline{P})$
1.	$P = \{s\},\ \overline{P} = \{b, c, d, t\}$	$4 + 5 = 9$
2.	$P = \{s, b\},\ \overline{P} = \{c, d, t\}$	$2 + 5 + 5 = 12$
3.	$P = \{s, c\},\ \overline{P} = \{b, d, t\}$	$4 + 5 + 6 + 4 = 19$
4.	$P = \{s, d\},\ \overline{P} = \{b, c, t\}$	$4 + 1 + 2 = 7$
5.	$P = \{s, b, c\},\ \overline{P} = \{d, t\}$	$5 + 2 + 4 = 11$
6.	$P = \{s, b, d\},\ \overline{P} = \{c, t\}$	$5 + 2 + 1 + 2 = 10$
7.	$P = \{s, c, d\},\ \overline{P} = \{b, t\}$	$4 + 6 + 4 + 2 = 16$
8.	$P = \{s, b, c, d\},\ \overline{P} = \{t\}$	$2 + 4 + 2 = 8$

The minimum capacity among these Cuts occurs for the Cut $(P, \overline{P})$ where $P = \{s, d\}, \overline{P} = \{b, c, t\}$. Therefore, according to *Max-Flow Min-Cut theorem*, the maximal flow, from source vertex s to sink vertex t, is 7 units.

Using Max-Flow Min-Cut Theorem, we can be able to determine the Maximal flow in the Transport Network from source s to sink t. But it does not explicitly provide the Maximal flow pattern. Moreover, it becomes cumbersome if the number of intermediate vertices between s and t is large. Therefore, we need an efficient method to find the Maximal Flow as well as Maximal Flow Pattern. Next, we present the Ford-Fulkerson method for solving Maximal Flow problem. The Ford-Fulkerson method is iterative. The Ford-Fulkerson method depends upon three important factors and these are relevant to many flow algorithms and problems: residual capacity, residual network, and augmenting path.

10.3 Residual Capacity and Residual Network

10.3.1 Residual Capacity

Given a Transport Network or Flow Network $G = (V, E)$ with source vertex s and sink vertex t. Consider a pair of vertices u, v in G. The amount of additional net flow, that can be transferred from u to v without exceeding the capacity $c(u, v)$, is called the *residual capacity* of the edge (u, v). It is denoted by $c_f(u, v)$ and given by

$$c_f(u, v) = c(u, v) - f(u, v)$$

Indeed, the residual capacity $c_f(u,v)$ of an edge $(u,\ v)$ must be always non-negative, i.e. $c_f(u,v) \geq 0$. When $c_f(u,v) = 0$, the corresponding edge $(u,\ v)$ is said to be saturated edge.

10.3.2 Residual Network

Given a Transport Network or Flow Network $G = (V,\ E)$ and a flow f. The *residual network* of G induced by f is $G_f = (V, E_f)$, where

$$E_f = \left\{ (u,v) \in V \times V | c_f(u,v) > 0 \text{ and } (u,v) \in E \right\}$$

Clearly,

$V(G) = V(G_f)$ and $E_f \subseteq E$. So, G_f is a spanning subgraph of G. Of course, the graph G is itself residual network, since it is spanning subgraph of itself.

Augmenting Path:

An *augmenting path* P is a simple path from source vertex s to sink vertex t in the residual network G_f.

Residual Capacity of Augmenting Path:

The maximum amount of net flow that can be transferred along the edges of an augmenting path P is called the *residual capacity of augmenting path* P. It is denoted by $c_f(P)$ and given by

$$c_f(P) = \min\left\{ c_f(u,v) | (u,v) \text{ is an edge on } P \right\}$$

10.4 Ford-Fulkerson Algorithm

Initial Step:

For each edge $(u,v) \in E(G)$,

Set $f(u,v) = 0$

Step 1:

While there exists an augmenting path P from source vertex s to sink vertex t in the residual network G_f then

Compute

$$c_f(P) = \min\left\{ c_f(u,v) | (u,v) \text{ is an edge on } P \right\}$$

Step 2:

For each edge $(u,\ v)$ on augmenting path P

Set $f(u,v) = f(u,v) + c_f(P)$

Step 3:

If there exists no an augmenting path P from source vertex s to sink vertex t in the residual network G_f, then Stop. Otherwise, go to Step 1.   □

At each iteration, we increase the flow value by finding an "augmenting path", along which we can send more flow and then augmenting the flow along this path. We repeat this procedure until no augmenting path can be found.

The value of the maximal flow will be

$$\sum c_f(P_i), \quad \text{for all } P_i \in \text{Maximal flow pattern}$$

where
Maximal flow pattern $= \{P_i | P_i$ is an augmenting path in residual network$\}$

10.5 Ford-Fulkerson Algorithm with Modification by Edmonds-Karp

10.5.1 Time Complexity of Ford-Fulkerson Algorithm

The Ford Fulkerson Method does not specify how to find the augmenting paths. For example, we can use either BFS or DFS to find the augmenting paths. In fact, without further knowledge on how to find the augmenting path, the best bound we have on the time complexity is $O(|E| * f^*)$, where $|E|$ is the number of edges in the graph and f^* is the maximum flow. This is based on the observation that it takes $O(|E|)$ time to find an augmenting path and every augmenting path increases the flow by at least 1. Finally, it should be noted that the Ford Fulkerson Method does not guarantee that it will terminate—there are some special cases involving irrational capacity where we will keep finding augmenting path with smaller capacities. However, if it does terminate (as it will in the case of integer capacities), it will return the correct answer. If we use BFS to find the augmenting path, then it is known as the *Edmonds-Karp Algorithm*.

10.5.2 Edmonds-Karp Algorithm

This algorithm is a variation on the Ford-Fulkerson method which is intended to increase the speed of the first algorithm. The idea is to try to choose good augmenting paths. In this algorithm, the augmenting path suggested is the augmenting path with the minimum number of edges [We can find this using Breadth First Search (BFS)]. The bound on Ford-Fulkerson method can be improved if we implement the computation of the augmenting path in Ford-Fulkerson method with BFS, i.e. if the augmenting path is a shortest path from source vertex s to sink vertex t in the residual network G_f. The total number of iterations of the algorithm using this strategy is $O(|V||E|)$. Thus, its running time is $O(|V||E|^2)$. we can actually prove that the algorithm will terminate in $O(|V||E|^2)$ time, a much tighter bound than what we have before.

Example 10.3 Using Ford-Fulkerson Algorithm, find the maximal flow of the following Transport Network in Fig. 10.5.

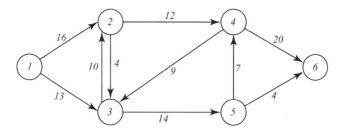

Fig. 10.5 A transport network

Solution:

Initially, we assign zero flow to every edge in the given Transport Network.

The first residual network is the original Transport Network itself. We need to find the augmenting path from source vertex 1 to sink vertex 6. According to Edmonds-karp algorithm, the augmenting path is the shortest path using BFS from 1 to 6 (Fig. 10.6).

The Breadth First Tree:

Fig. 10.6 Breadth First Tree

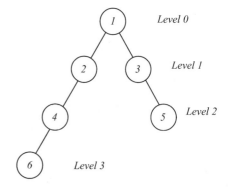

Therefore, the augmenting path, which is the shortest path obtained by BFS, is P_1: 1–2–4–6. The residual capacity of this path is $c_f(P_1) = 12$. Augmenting the above path by pushing 12 units of flow along the path, the following augmented network and the corresponding residual network have been obtained in Fig. 10.7. In residual network, the residual capacity of an edge from node u to node v is simply its unused capacity which is the difference between its capacity and its flow. It is shown in the direction of the edge. In the opposite direction from v to u, the value indicates the flow of the edge (Fig. 10.7).

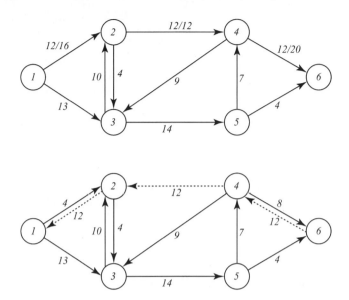

Fig. 10.7 Augmented network and its corresponding residual network

Again, we find the augmenting path from source vertex 1 to sink vertex 6 in the residual network of Fig. 10.7. According to Edmonds-Karp algorithm, the augmenting path is the shortest path using BFS from 1 to 6.

The Breadth First Tree:

Fig. 10.8 Breadth First Tree

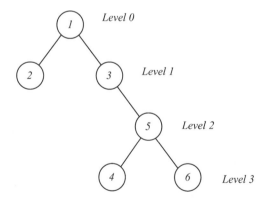

Here, the augmenting path, which is the shortest path obtained by BFS, is P_2: 1–3–5–6. The residual capacity of this path is $c_f(P_2) = 4$. So, pushing 4 units of flow along the path, the following augmented network and the corresponding residual network have been obtained in Fig. 10.9.

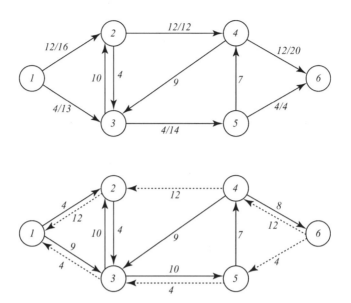

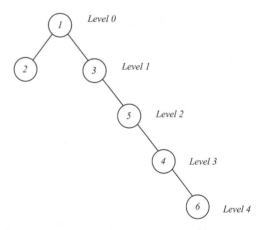

Fig. 10.9 Augmented network and its corresponding residual network

Next, we again find the augmenting path using BFS from 1 to 6 in the residual network of Fig. 10.9. The following Fig. 10.10 shows the Breadth First Tree obtain by BFS.

The Breadth First Tree:

Fig. 10.10 Breadth First
Tree

Now, the augmenting path, which is the shortest path obtained by BFS, is P_3: 1–3–5–4–6. The residual capacity of this path is $c_f(P_3) = 7$. After, pushing 7 units of flow along the path, the following augmented network and the corresponding residual network have been obtained in Fig. 10.11.

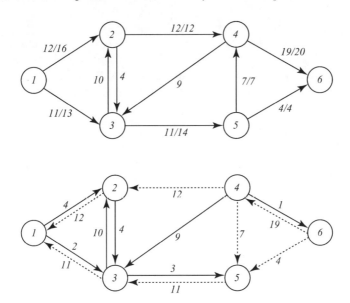

Fig. 10.11 Augmented network and its corresponding residual network

At this stage, there are no more shortest paths from 1 to 6. So, there exists no augmenting path from 1 to 6 in the residual network of Fig. 10.11. Consequently, the algorithm stops executing.

Therefore, the Maximal flow is $c_f(P_1) + c_f(P_2) + c_f(P_3) = 12 + 4 + 7 = 23$ units. The corresponding Maximal flow pattern is $\{P_1, P_2, P_3\}$. □

Now, the above value for Maximal flow can be verified in the following way. Let us consider the set of vertices that are reachable from source 1 in the residual network of Fig. 10.11. This set includes vertices 2, 3 and 5. Now, we consider the cut $(P, \overline{P})$, where $P = \{1, 2, 3, 5\}$, P contains those vertices which are reachable from source 1 and $\overline{P} = \{4, 6\}$. From Fig. 10.5, the capacity of this cut is $12 + 7 + 4 = 23$, which is the required Maximal Flow. Hence, it is verified.

Example 10.4 Using Ford-Fulkerson Algorithm, find the maximal flow of the following Transport Network in Fig. 10.12.

Fig. 10.12 A transport network

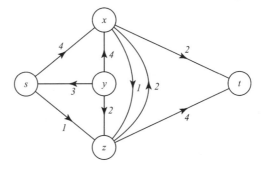

Solution:
Initially, we assign zero flow to all edges in the given Transport Network.

The first residual network is the original Transport Network itself. We need to find the augmenting path from source vertex s to sink vertex t. According to Edmonds-karp algorithm, the augmenting path is the shortest path using BFS from s to t (Fig. 10.13).

The Breadth First Tree:

Fig. 10.13 Breadth First
Tree

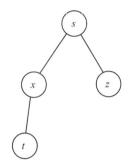

Therefore, the augmenting path, which is the shortest path obtained by BFS, is $P_1 : s - x - t$. The residual capacity of this path is $c_f(P_1) = 2$. Augmenting the above path by pushing 2 units of flow along the path, the following augmented network and the corresponding residual network have been obtained in Fig. 10.14. In residual network, the residual capacity of an edge from node u to node v is

Fig. 10.14 Augmented
network and its
corresponding residual
network

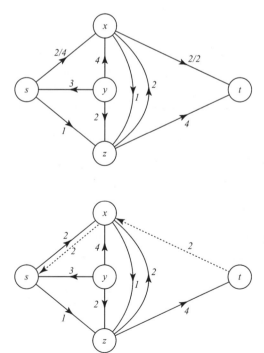

shown in the direction of the edge. In the opposite direction from v to u, the value indicates the flow of the edge (Fig. 10.14).

Again, we find the augmenting path from source vertex s to sink vertex t in the residual network of Fig. 10.14.

The Breadth First Tree:

Fig. 10.15 Breadth First Tree

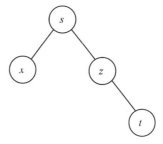

Here, the augmenting path, which is the shortest path obtained by BFS, is $P_2 : s - z - t$. The residual capacity of this path is $c_f(P_2) = 1$. So, pushing 1 unit of flow along the path, the following augmented network and the corresponding residual network have been obtained in Fig. 10.16.

Fig. 10.16 Augmented network and its corresponding residual network

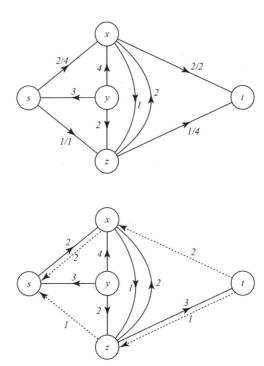

Next, we again find the augmenting path using BFS from s to t in the residual network of Fig. 10.16. The following Fig. 10.17 shows the Breadth First Tree obtain by BFS.

The Breadth First Tree:

Fig. 10.17 Breadth First
Tree

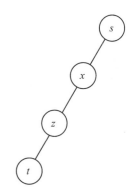

Now, the augmenting path, which is the shortest path obtained by BFS, is $P_3 : s - x - z - t$. The residual capacity of this path is $c_f(P_3) = 1$. After, adding 1 unit of flow along the path, the following augmented network and the corresponding residual network have been obtained in Fig. 10.18.

Fig. 10.18 Augmented
network and its
corresponding residual
network

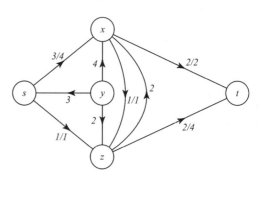

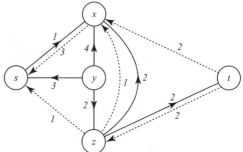

At this stage, there are no more shortest paths from s to t. So, there exists no augmenting path from s to t in the residual network of Fig. 10.18. Consequently, the algorithm halts.

Therefore, the Maximal flow is $c_f(P_1) + c_f(P_2) + c_f(P_3) = 2 + 1 + 1 = 4$ units. The corresponding Maximal flow pattern is $\{P_1, P_2, P_3\}$.   □

Let us consider the set of vertices that are reachable from source s in the residual network of Fig. 10.18. This set includes only vertex x. Now, we consider the cut $(P, \overline{P})$, where $P = \{s, x\}$, P contains those vertices which are reachable from source s and $\overline{P} = \{y, z, t\}$. From Fig. 10.12, the capacity of this cut is $2 + 1 + 1 = 4$, which is the required Maximal Flow obtained in example 10.1. Hence, it is verified with the Max-Flow Min-Cut result.

10.6 Maximal Flow: Applications

10.6.1 Multiple Sources and Sinks

This can be converted to a single source, single sink situation, so far discussed above, by introducing

1. a new vertex s' (super source or dummy source) from which there is an edge to each source, capacity of the new edges is ∞.
2. a new vertex t' (super sink or dummy sink). There is an edge from each target to t', capacity of the new edges is ∞ (Fig. 10.19).

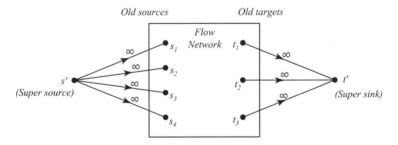

Fig. 10.19 Transport network with multi-source and multi-sink

10.6.2 Maximum Bipartite Matching

Consider the problem, we have n people and m jobs. We know what jobs can be done by which person.

Our problem is to find a job assignment of at most one job to one person so that the maximal number of jobs done. This can be represented by a bipartite graph.

Bipartite Graph:

- Suppose we have a set of people $L = \{p_1, p_2, p_3, p_4\}$ and set of jobs $R = \{j_1, j_2, j_3, j_4, j_5\}$.
- Each person can do only some of the jobs.
- This problem can be model in the following bipartite graph as shown in Fig. 10.20a.

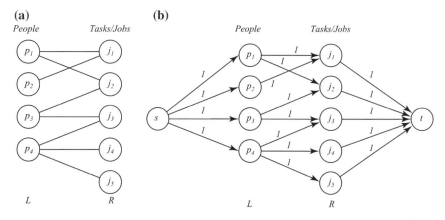

Fig. 10.20 (a) Bipartite graph showing Bipartite matching, (b) Bipartite matching transferred into maximal network flow problem

An Optimal Job Assignment is equivalent to Maximum Bipartite Matching problem.

Bipartite Matching:

(i) A matching gives an assignment of people to tasks.

(ii) We wish to get as many tasks done as possible.

(iii) So, we need a maximum matching: one that contains as many edges as possible.

This problem can be converted into a maximal flow problem.

We introduce a source connected to all persons and a sink connected to all jobs with all capacities 1, shown in Fig. 10.20b.

There is an integral solution of the flow giving maximal matching.

Exercises:

1. Apply Ford-Fulkerson Algorithm with modification by Edmonds-Karp to find the maximal flow for the following network in Fig. 10.21.

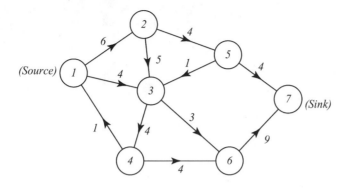

Fig. 10.21

2. Apply Edmonds-Karp Algorithm to find the maximal flow from source s to sink t for the following networks (Fig. 10.22).

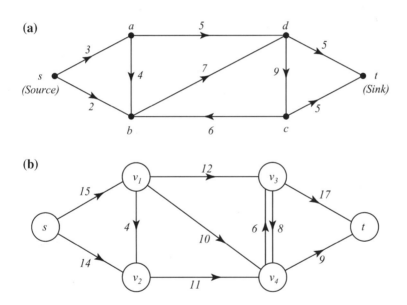

Fig. 10.22

3. Use Ford-Fulkerson Algorithm to find the maximal flow from source s to sink t for the following network. Find a cut with capacity equal to this maximal flow (Fig. 10.23).

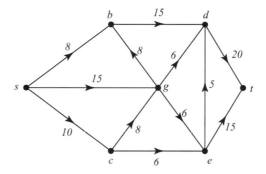

Fig. 10.23

4. Find all possible $(P, \overline{P})$ cuts in the transport network shown in Fig. 10.22a. Hence, determine the maximal flow using Max-Flow Min-Cut Theorem.
5. Show that a maximum flow in a network $G = (V, E)$ can always be found by a sequence of at most $|E|$ augmenting paths.

Appendix

C++ Program 1: The Dijkstra's Algorithm has been implemented in the following C++ program.

DIJKSTRA.CPP

```
#include<iostream.h>
#include<fstream.h>
#include<conio.h>
#include<math.h>
#define START 1  // Starting vertex
#define END 6    // Ending vertex
#define P 1   // Permanent
#define T 0   // Temporary
#define INFINITY 9999

ifstream in("dijk_in.txt");
ofstream out("Dij_Out.txt");

struct node
{
double label;     // label of the vertex
int status;       // status of the vertex P or T
int pred;         // Predecessor of the vertex
}
vertex[50];       // vertex array with three attributes
double a[50][50];  // Weight matrix for the edges
int n;            // number of vertices
int e;            // number of edges
main()
{
```

```
int i,j,u,v;
int path[50];
double d;
double dijkstra(int s,int t);
void short_path(int v);
clrscr();
in>>n>>e;
 out<<"The number of vertices: "<<n<<endl;
 out<<"The number of edges: "<<e<<endl;
for(u=0;u<=n;u++)
 for(v=0;v<=n;v++)
   a[u][v]=INFINITY;
do
{
in>>u>>v>>d;
a[u][v]=d;
}
while(!in.eof());
in.close();
out<<"\nThe Adjacency matrix(Weight matrix):"<<endl;
   for(i=1;i<=n;i++)
{
   for(j=1;j<=n;j++)
        out<<a[i][j]<<" ";
        out<<endl;
        }
if(dijkstra(START,END))
{
out<<"\nThe shortest distance from "<<START<<" to "<<END<<" :
"<<dijkstra(START,END)<<endl;
out<<"\nThe shortest path : ";
   short_path(END);
}
else
out<<"There is no path:"<<"\n";
cin.get();
return 0;
}
double dijkstra(int s,int t)
{
int i,j,k;
double min;
for(j=1;j<=n;j++)
{
vertex[j].label=INFINITY;
```

```
vertex[j].status=T;
 }
vertex[s].label=0;
vertex[s].status=P;
vertex[s].pred=0;
k=s;
do
{
for(j=1;j<=n;j++)
{
if((vertex[j].status==T)&&(a[k][j]!=INFINITY))
{
  if(vertex[j].label>vertex[k].label+a[k][j])
{
    vertex[j].label=vertex[k].label+a[k][j];
    vertex[j].pred=k;
          }
      }
    }
  k=0;
  min=INFINITY;
  for(i=1;i<=n;i++)
{
   if((vertex[i].status==T)&&(vertex[i].label<min))
{
          min=vertex[i].label;
          k=i;
          }
    }
          vertex[k].status=P;
  if(k==0)
    return(0);
}
while(k!=t);
  return(vertex[t].label);
 }


  void short_path(int v)
{
  int a,i,u,l=0;
  int pred[50];
  static int path[50];   // Shortest Path array
  u=v;
  for(u=END;u!=0;u=vertex[u].pred)
```

```
       path[++l]=u;
   for(i=l;i>1;i--)
   out<<path[i]<<"->";
     out<<END;
   }
```

Example 1: Consider the network as shown in Fig. 5.1. In the above program, the input stream read data from input file "dijk_in.txt" and output stream write data to a output file "DIJ_OUT.TXT". The first line of the input file "dijk_in.txt" is the number of vertices and edges. The rest is the adjacency matrix or distance matrix of the network shown in Fig. 5.1.

dijk_in.txt

```
6 9
1 2 18
1 4 15
2 3 9
3 6 28
4 2 6
4 3 14
4 5 7
5 3 10
5 6 36
```

After the execution of the program "DIJKSTRA.CPP", the following output file "DIJ_OUT.TXT" is generated. The output shows the shortest route from the starting vertex 1 to the ending vertex 6.

DIJ_OUT.TXT

The number of vertices: 6
The number of edges: 9

The Adjacency matrix(Weight matrix):
9999 18 9999 15 9999 9999
9999 9999 9 9999 9999 9999
9999 9999 9999 9999 9999 28
9999 6 14 9999 7 9999
9999 9999 10 9999 9999 36
9999 9999 9999 9999 9999 9999

The shortest distance from 1 to 6 : 55

The shortest path : 1->2->3->6

C++ Program 2: The Floyd's Algorithm has been implemented in the following C++ program.

FLOYD.CPP


```cpp
#include <iostream.h>
#include <fstream.h>
#include <conio.h>
#include <stdio.h>

ifstream in("floyd.txt");
ofstream out("fld_out.txt");
float d[50][50];    // Distance Matrix
int s[50][50];       // Node Sequence Matrix
main()
{
float dist;
int n;  // number of vertices
int u,v,i,j,k;
char ch;
void floyd(int n);
void short_path(int,int);
clrscr();
 in>>n;
 out<<n<<endl;
for(i=1;i<=n;i++)
  for(j=1;j<=n;j++)
    s[i][j]=0;
    for(i=1;i<=n;i++)
          for(j=1;j<=n;j++)
            if(i!=j)
             d[i][j]=9999;
            else
             d[i][j]=0;
    while(!in.eof())
            {
          in>>i>>j>>dist;
          d[i][j]=dist;
            }
```

```
            in.close();
    out<<"\nThe Adjacency matrix(Weight matrix):"<<endl;
      for(i=1;i<=n;i++)
{

      for(j=1;j<=n;j++)
            out<<d[i][j]<<"  ";
            out<<endl;
            }
floyd(n);
do
{
cout<<"Enter the vertices:";
cin>>u>>v;
out<<"\nThe Shortest Distance from "<<u<<" to "<<v<<": "<<d[u][v]<<endl;
out<<"\nThe Shortest Path: ";
out<<u;
short_path(u,v);
out<<"->"<<v<<endl;
cout<<"\nPress ENTER to continue..."<<endl;
cout<<"\nOtherwise press any key to quit..."<<endl;
cin.sync();
}
while(cin.get()=='\n');
return 0;
}

void floyd(int n)
{
int i,j,k;
   for(k=1;k<=n;k++)
{
    out<<"---------------------------------------"<<endl;
    out<<"Step:"<<k<<endl;
    for(i=1;i<=n;i++)
    for(j=1;j<=n;j++)
    if(i!=k && j!=k && i!=j)
          if(d[i][j]>d[i][k]+d[k][j])
          {
             d[i][j]=d[i][k]+d[k][j];
             s[i][j]=k;
                  }
    out<<"\nDistance Matrix"<<endl;
      for(i=1;i<=n;i++)
{
```

```
        for(j=1;j<=n;j++)
         out<<d[i][j]<<" ";
        out<<endl;
          }
   out<<"\nNode Sequence Matrix"<<endl;
    for(i=1;i<=n;i++)
{
        for(j=1;j<=n;j++)
         out<<s[i][j]<<" ";
        out<<endl;
          }
         } //end of k loop
         out<<"----------------------------------------"<<endl;
    }

void short_path(int i,int j)
{
int k;
k=s[i][j];
if(k!=0)
{
  short_path(i,k);
  out<<"->"<<k;
  short_path(k,j);
    }
}
```

Example 2: Consider the network as shown in Fig. 5.6. In the above program, the input stream is used to read data from input file "floyd.txt" and output stream is used to write data to a output file "FLD_OUT.TXT". The first line of the input file "floyd.txt" is the number of vertices. The rest is the adjacency matrix or distance matrix of the network shown in Fig. 5.6.

floyd.txt

```
5
1 2 8
2 1 8
2 5 5
5 2 5
3 2 1
2 3 2
1 3 3
1 4 5
```

4 1 6
3 4 3
4 5 7
3 5 4

After the execution of the program "FLOYD.CPP", the following output file "FLD_OUT.TXT" is generated. The output shows the shortest routes from node 1 to node 5, node 4 to node 2, node 5 to node 4 and node 4 to node 3.

FLD_OUT.TXT

5

The Adjacency matrix(Weight matrix):
0 8 3 5 9999
8 0 2 9999 5
9999 1 0 3 4
6 9999 9999 0 7
9999 5 9999 9999 0

Step:1

Distance Matrix
0 8 3 5 9999
8 0 2 13 5
9999 1 0 3 4
6 14 9 0 7
9999 5 9999 9999 0

Node Sequence Matrix
0 0 0 0 0
0 0 0 1 0
0 0 0 0 0
0 1 1 0 0
0 0 0 0 0

Step:2

Distance Matrix
0 8 3 5 13
8 0 2 13 5
9 1 0 3 4
6 14 9 0 7
13 5 7 18 0

Node Sequence Matrix
0 0 0 0 2
0 0 0 1 0
2 0 0 0 0
0 1 1 0 0
2 0 2 2 0

Step:3

Distance Matrix
0 4 3 5 7
8 0 2 5 5
9 1 0 3 4
6 10 9 0 7
13 5 7 10 0

Node Sequence Matrix
0 3 0 0 3
0 0 0 3 0
2 0 0 0 0
0 3 1 0 0
2 0 2 3 0

Step:4

Distance Matrix
0 4 3 5 7
8 0 2 5 5
9 1 0 3 4
6 10 9 0 7
13 5 7 10 0

Node Sequence Matrix
0 3 0 0 3
0 0 0 3 0
2 0 0 0 0
0 3 1 0 0
2 0 2 3 0

Step:5

Distance Matrix
0 4 3 5 7
8 0 2 5 5

9 1 0 3 4
6 10 9 0 7
13 5 7 10 0

Node Sequence Matrix
0 3 0 0 3
0 0 0 3 0
2 0 0 0 0
0 3 1 0 0
2 0 2 3 0

The Shortest Distance from 1 to 5: 7

The Shortest Path: 1->3->5

The Shortest Distance from 4 to 2: 10

The Shortest Path: 4->1->3->2

The Shortest Distance from 5 to 4: 10

The Shortest Path: 5->2->3->4

The Shortest Distance from 4 to 3: 9

The Shortest Path: 4->1->3

C++ Program 3: The Breadth First Search Algorithm has been implemented in the following C++ program. This program uses two data structures to implement the Breadth First Traversal: a color marker for each vertex and a queue. In the beginning all vertices are coloured white. White vertices are undiscovered vertices not yet in the queue. We will colour the vertices gray when we enqueue(add to the end of the queue) them. The gray vertices are discovered but have undiscovered adjacent vertices. We will colour the vertices black when we dequeue(remove from the front of the queue) them. The black vertices are discovered and are adjacent to only other black or gray vertices. While the queue is not empty, we run the loop of the Breadth First Search. The algorithm proceeds by removing a vertex u from the queue and examining each out-edge (u,v). If an adjacent vertex v is not already discovered, it is colored gray and placed in the queue. After all of the out-edges are examined, vertex u is colored black and deleted from the queue. This process is repeated. After the loop has finished, all nodes reachable from the starting vertex are black. The unreachable vertices are still white.

BFS.CPP

```cpp
#include <iostream.h>
#include <fstream.h>
#include <conio.h>

//Basic Definitions

#define WHITE 0
#define GRAY 1
#define BLACK 2
#define MAX_NODES 100
#define INFINITY 9999
#define START 1
#define END 6

//Declarations
int n; // number of nodes
int e; // number of edges
int ad[MAX_NODES][MAX_NODES]; // adjacency matrix
int color[MAX_NODES]; // needed for breadth-first search
int pred[MAX_NODES]; // array to store shortest path
int k; // level
    ifstream in("BFS.TXT");
    ofstream out("BFS_OUT.TXT");

//A Queue for Breadth-First Search

int head,tail;
int q[MAX_NODES+2];
void bfs (int start, int target);
void enqueue (int x)
{
    q[tail] = x;
    tail++;
}

int dequeue ()
{
    int x = q[head];
    head++;
    return x;
}

//Breadth-First Search
```

```
void bfs (int start, int target)
{
 void short_path(int,int);
   int u,v;
   for (u=1; u<=n; u++)
{
        color[u] = WHITE;
   }
   head = tail = 0;
   k=0;
   enqueue(start);
   pred[start] = -1;
   while (head!=tail)
{
        u = dequeue();
        // Search all adjacent white nodes v.
        // enqueue v.
        for (v=1; v<=n; v++)
{
           if (color[v]==WHITE && ad[u][v]>0)
{
                color[v] = GRAY;
                pred[v] = u;
                enqueue(v);
                         }
                  }
           color[u]=BLACK;
           out<<u<<" ";
                  }
           out<<"\n";
   // If the color of the target node is black now,
   // it means that we reached it.
   if(color[target]==BLACK)
{
     short_path(END,0);
     out<<endl;
           }
}


//Reading the input file and the main program

void read_input_file()
{
```

```cpp
   int a,b,c,i,j;

   // read number of nodes and edges
   in>>n>>e;
   out<<"The number of nodes: "<<n<<endl;
   out<<"\nThe number of edges: "<<e<<endl;


   // initialize empty adjacency matrix
   for (i=1; i<=n; i++)
{
        for (j=1; j<=n; j++)
{
            ad[i][j] = 0;
        }
   }

   // read adjacency matrix
   while(!in.eof())
{
   for (i=0; i<e; i++)
{
   in>>a>>b>>c;
        ad[a][b] = c;
        }
   }
   in.close();
   out<<"\nThe Adjacency matrix:"<<endl;
   for(i=1;i<=n;i++)
{
   for(j=1;j<=n;j++)
        out<<ad[i][j]<<" ";
        out<<endl;
        }
}


int main ()
{
   clrscr();
   read_input_file();
   out<<"\n"<<"The Breadth First Traversal: ";
   bfs(START,END);
   getch();
   return 0;
}
```

```
void short_path(int t,int l)
{
 int i,u,v;
 static int path[50];
 v=t;
 u=pred[v];
 path[l++]=u;
 if(u!=START)
 short_path(u,l);
 else
   {
   out<<"\nThe shortest Distance: "<<l<<endl;
   out<<"\nThe shortest Path: ";
 for(i=l-1;i>=0;i--)
 out<<path[i]<<"->";
 out<<END;
   }
}
```

Example 3: Consider the graph as shown in Fig. 5.15. In the above program, the input stream read data from input file "bfs.txt" and output stream write data to a output file "BFS_OUT.TXT". The first line of the input file "bfs.txt" is the number of vertices and edges. The rest is the adjacency matrix of the graph shown in Fig. 5.15.

bfs.txt

8 10
1 2 1
2 1 1
2 3 1
3 2 1
2 4 1
4 2 1
3 4 1
4 3 1
3 5 1
5 3 1
4 5 1
5 4 1
4 6 1
6 4 1

4 7 1
7 4 1
6 8 1
8 6 1
1 8 1
8 1 1

After the execution of the program "BFS.CPP", the following output file "BFS_OUT. TXT" is produced. The following output shows the shortest path as well as distance from vertex 1 to vertex 6.

BFS_OUT.TXT

The number of nodes: 8

The number of edges: 10

The Adjacency matrix:
0 1 0 0 0 0 0 1
1 0 1 1 0 0 0 0
0 1 0 1 1 0 0 0
0 1 1 0 1 1 1 0
0 0 1 1 0 0 0 0
0 0 0 1 0 0 0 1
0 0 0 1 0 0 0 0
1 0 0 0 0 1 0 0

The Breadth First Traversal: 1 2 8 3 4 6 5 7

The shortest Distance: 2

The shortest Path: 1->8->6

C++ Program 4: The Prim's Algorithm has been also implemented in the following C++ program. Prim's algorithm is a greedy algorithm that finds the minimum spanning tree of a graph.

PRIM.CPP

```
#include<iostream.h>
#include<fstream.h>
#include<conio.h>
ifstream in("Prim.txt");
ofstream out("Prim_Out.txt");
int Prim(int wt[50][50],int n)
            {
int visited[50]={0},min=9999,minwt=0;
int a,b,u,v,i,j,w;
int ne=0;   //number of edges in minimum spanning tree
   visited[1]=1;
   while(ne!=n-1)
  {
    for(i=1,min=9999;i<=n;i++)
        for(j=1;j<=n;j++)
          if(wt[i][j]<min)
            if(visited[i]==1 && visited[j]==0)
            {
            min=wt[i][j];
            a=u=i;
            b=v=j;
            }
    if(visited[u]==0 || visited[v]==0)
    {
    out<<"\n"<<++ne<<". Edge"<<"  ("<<a<<","<<b<<") "<<"Weight "<<min<<endl;
    minwt+=min;
    visited[b]=1;
    }
    }
    return(minwt);
}


void main()
{
int Prim(int wt[50][50],int n);
int i,j;
int n;        // number of vertices
int wt[50][50];  // weight of the edges
int w;
int minwt;      // minimum weight
clrscr();
in>>n;
```

```
out<<"The number of Vertices:"<<n<<endl;
for(i=1;i<=n;i++)
 for(j=1;j<=n;j++)
    wt[i][j]=9999;
    while(!in.eof())
    {
      in>>i>>j>>w;
      if(w==0)
      wt[i][j]=9999;
      else
      wt[i][j]=w;
      }
    in.close();
          out<<"\nThe Adjacency matrix(Weight matrix):"<<endl;
    for(i=1;i<=n;i++)
{
    for(j=1;j<=n;j++)
          out<<wt[i][j]<<" ";
          out<<endl;
          }
          out<<"\nThe Spanning Tree contains the following Edges:"<<endl;
    minwt=Prim(wt,n);
    out<<"\nMinimum Weight="<<minwt<<endl;
    cin.get();
    }
```

Example 4: Consider the graph as shown in Fig. 5.11. In the above program, the input stream read data from input file "prim.txt" and output stream write data to a output file "PRIM_OUT.TXT". The first line of the input file "prim.txt" is the number of vertices. The rest is the adjacency matrix or weight matrix of the graph shown in Fig. 5.11.

prim.txt

6
1 1 0
1 2 2
1 3 4
2 1 2
2 2 0
2 3 7
2 4 11
3 1 4

3 2 7
3 3 0
3 4 8
3 6 1
4 2 11
4 3 8
4 4 0
4 5 6
5 4 6
5 5 0
5 6 9
6 3 1
6 5 9
6 6 0

After the execution of the program "PRIM.CPP", the following output file "PRIM_OUT.TXT" is produced. The following output shows the Minimum Spanning Tree with Minimum Weight.

PRIM_OUT.TXT

The number of Vertices:6

The Adjacency matrix(Weight matrix):
9999 2 4 9999 9999 9999
2 9999 7 11 9999 9999
4 7 9999 8 9999 1
9999 11 8 9999 6 9999
9999 9999 9999 6 9999 9
9999 9999 1 9999 9 9999

The Spanning Tree contains the following Edges:

1. Edge (1,2) Weight 2

2. Edge (1,3) Weight 4

3. Edge (3,6) Weight 1

4. Edge (3,4) Weight 8

5. Edge (4,5) Weight 6

Minimum Weight=21

C++ Program 5: The Kruskal's Algorithm has been also implemented in the following C++ program. This program of Kruskal's algorithm for finding a minimum spanning tree uses a data structure for maintaining a collection of disjoint sets. It supports the following three operations:
- makeset(x) - create a new set containing the single element x.
- merge(x,y) - replace the two sets containing x and y by their union.
- find(x) - returns the representative of the set containing x.

KRUSKAL.CPP

```
#include<iostream.h>
#include<fstream.h>
#include<conio.h>
#define MAX 100
ifstream in("krus_in.txt");
ofstream out("Krus_Out.txt");

struct edge_info
 {
   int u, v, weight;
 }
edge[MAX];
int wt[MAX][MAX];
int tree[MAX][3], set[MAX];
int n;
int readedges();
void makeset();
int find(int);
void merge(int, int);
void arrange_edges(int);
void spanningtree(int);
int readedges()
{
        int i, j, k, w;
        k = 1;
        out << "\nThe number of Vertices in the Graph : ";
        in>>n;
        out<<n<<endl;
    for(i=1;i<=n;i++)
    for(j=1;j<=n;j++)
        wt[i][j]=9999;
    while(!in.eof())
```

```
{
    in>>i>>j>>w;
    edge[k].u = i;
    edge[k].v = j;
    if(w==0)
    {
    wt[i][j]=9999;
         }
    else
         {
    if(j>i)
    edge[k++].weight=w;
    wt[i][j]=w;
         }
    }
    in.close();
    out<<"\nThe Adjacency matrix(Weight matrix):\n"<<endl;
    for(i=1;i<=n;i++)
{
    for(j=1;j<=n;j++)
         out<<wt[i][j]<<" ";
         out<<endl;
         }
         return (k - 1);
}


void makeset()
{
         int i;
         for (i = 1; i <= n; i++)
                  set[i] = i;
}


int find(int vertex)
{
         return (set[vertex]);
}


void merge(int v1, int v2)
{
         int i, j;
         if (v1 < v2)
```

```
                    set[v2] = v1;
        else
                    set[v1] = v2;
}

// sort set of edges in non-decreasing order by weight(applying bubblesort)

void arrange_edges(int k)
{
        int i, j;
        struct edge_info temp;
        for (i = 1; i < k; i++)
                    for (j = 1; j <= k - i; j++)
                                if (edge[j].weight > edge[j + 1].weight)
                                {
                                            temp = edge[j];
                                            edge[j] = edge[j + 1];
                                            edge[j + 1] = temp;

                                }

}


void spanningtree(int k)
{
        int i, t, sum;
        arrange_edges(k);
        t = 1;
        sum = 0;
        out<<"\nThe sorted set of edges in non-decreasing order by weight(after
applying bubblesort)"<<endl;
        for (i=1;i<=k;i++)
{

        out<<edge[i].u<<" "<<edge[i].v<<" "<<edge[i].weight<<endl;
        }
        cin.get();
        for (i = 1; i < k; i++)
                    if (find (edge[i].u) != find (edge[i].v))
                    {
                                tree[t][1] = edge[i].u;
                                tree[t][2] = edge[i].v;
                                tree[t][3] = edge[i].weight;
                                merge(edge[t].u, edge[t].v);
                                t++;

                    }
        out << "\nThe Edges of the Minimum Spanning Tree are\n\n";
```

```
            for (i = 1; i < n; i++)
{
                    out <<i<<". "<<"Edge "<<tree[i][1] << " - " << tree[i][2] <<"
Weight: "<<tree[i][3]<<endl;
                    sum+=tree[i][3];
                    }
                    out << "\nThe Weight of the Minimum Spanning Tree is : " <<
sum;
}




int main()
{
        int num_edge;    /* number of edges in minimum spanning tree */
    int min_weight;   /* weight of minimal spanning tree */
        clrscr();
        num_edge = readedges();
        makeset();
        spanningtree(num_edge);
        return 0;
}
```

Example 5: Consider the graph as shown in Fig. 5.11. In the above program, the input stream read data from input file "krus_in.txt" and output stream write data to a output file "KRUS_OUT.TXT". The first line of the input file "krus_in.txt" is the number of vertices. The rest is the adjacency matrix or weight matrix of the graph shown in Fig. 5.11.

krus_in.txt

```
6
1 1 0
1 2 2
1 3 4
2 1 2
2 2 0
2 3 7
2 4 11
3 1 4
3 2 7
3 3 0
3 4 8
3 6 1
4 2 11
```

4 3 8
4 4 0
4 5 6
5 4 6
5 5 0
5 6 9
6 3 1
6 5 9
6 6 0

After the execution of the program "KRUSKAL.CPP", the following output file "KRUS_OUT.TXT" is produced. The following output shows the Minimum Spanning Tree with Minimum Weight.

KRUS_OUT.TXT

The number of Vertices in the Graph : 6

The Adjacency matrix(Weight matrix):

```
9999 2 4 9999 9999 9999
2 9999 7 11 9999 9999
4 7 9999 8 9999 1
9999 11 8 9999 6 9999
9999 9999 9999 6 9999 9
9999 9999 1 9999 9 9999
```

The sorted set of edges in non-decreasing order by weight(after applying bubblesort)
3 6 1
1 2 2
1 3 4
4 5 6
2 3 7
3 4 8
5 6 9
2 4 11

The Edges of the Minimum Spanning Tree are

1. Edge 3 - 6 Weight: 1
2. Edge 1 - 2 Weight: 2
3. Edge 1 - 3 Weight: 4
4. Edge 4 - 5 Weight: 6
5. Edge 3 - 4 Weight: 8

The Weight of the Minimum Spanning Tree is : 21

C++ Program 6: The Ford-Fulkerson's (with modification by Edmonds-Karp) Algorithm has been implemented in the following C++ program.

MAXFLOW.CPP

```cpp
//The Ford-Fulkerson Algorithm in C++
#include <iostream.h>
#include <fstream.h>
#include <conio.h>
//Basic Definitions
#define WHITE 0
#define GRAY 1
#define BLACK 2
#define MAX_NODES 100
#define INFINITY 9999
#define START 1
#define END 6
//Declarations
int n; // number of nodes
int e; // number of edges
int capacity[MAX_NODES][MAX_NODES]; // capacity matrix
int flow[MAX_NODES][MAX_NODES];    // flow matrix
int color[MAX_NODES]; // needed for breadth-first search
int pred[MAX_NODES]; // array to store augmenting path
int k=0; // number of augmenting paths
   ifstream in("MF.TXT");
   ofstream out("MF_OUT.TXT");
int min (int x, int y) {
   return x<y ? x : y; // returns minimum of x and y
}
```

//A Queue for Breadth-First Search

```cpp
int head,tail;
int q[MAX_NODES+2];

void enqueue (int x)
{
   q[tail] = x;
   tail++;
}
```

```
int dequeue ()
{
   int x = q[head];
   head++;
   return x;
}


//Breadth-First Search for an augmenting path

int bfs (int start, int target)
{
  void short_path(int,int);
   int u,v;
   for (u=1; u<=n; u++)
{
          color[u] = WHITE;
   }
   head = tail = 0;
   enqueue(start);
   pred[start] = -1;
   while (head!=tail)
{
          u = dequeue();
          // Search all adjacent white nodes v. If the capacity
          // from u to v in the residual network is positive,
          // enqueue v.
          for (v=1; v<=n; v++)
{
              if (color[v]==WHITE && capacity[u][v]-flow[u][v]>0)
{
                     color[v] = GRAY;
                     pred[v] = u;
                     enqueue(v);
                               }
                     }
                color[u]=BLACK;
                          }

   // If the color of the target node is black now,
   // it means that we reached it.
   if(color[target]==BLACK)
{
   out<<"\n"<<++k<<"."<<"The Augmenting Path: ";
   short_path(END,0);
```

```
        out<<endl;
                }
        return color[target]==BLACK;
}


//Ford-Fulkerson Algorithm

int max_flow (int source, int sink)
{
    int i,j,u;
    // Initialize empty flow.
    int max_flow = 0;
    for (i=1; i<=n; i++)
{
            for (j=1; j<=n; j++)
{
                flow[i][j] = 0;
                }
        }

    out<<"\nThe Maximal Flow Pattern is the collection of all Augmenting
Paths:"<<endl;

    // While there exists an augmenting path,
    // increment the flow along this path.
    while (bfs(source,sink))
{
            // Determine the amount by which we can increment the flow.
            int increment = INFINITY;
            for (u=n; pred[u]>=1; u=pred[u])
{
                increment = min(increment,capacity[pred[u]][u]-flow[pred[u]][u]);
                }
            out<<"\n The Residual Capacity of the Augmenting Path:
"<<increment<<endl;
            // Now increment the flow.
            for (u=n; pred[u]>=1; u=pred[u])
{

                flow[pred[u]][u] += increment;
                flow[u][pred[u]] -= increment;
                                }
            max_flow += increment;
        }
    // No augmenting path anymore. Stop.
```

```
    return max_flow;
}


//Reading the input file and the main program
void read_input_file()
{
    int a,b,c,i,j;

    // read number of nodes and edges
    in>>n>>e;
    out<<"The number of nodes: "<<n<<endl;
    out<<"\nThe number of edges: "<<e<<endl;

    // initialize empty capacity matrix
    for (i=1; i<=n; i++)
{
        for (j=1; j<=n; j++)
{
            capacity[i][j] = 0;
        }
    }
    // read edge capacities
    while(!in.eof())
{
    for (i=0; i<e; i++)
{
    in>>a>>b>>c;
            capacity[a][b] = c;
        }
    }
    in.close();
    out<<"\nThe Adjacency matrix(Capacity matrix):"<<endl;
    for(i=1;i<=n;i++)
{
    for(j=1;j<=n;j++)
        out<<capacity[i][j]<<" ";
        out<<endl;
        }
}

int main ()
{
    clrscr();
```

```
    read_input_file();
    out<<"\n"<<"The Maximal flow: "<<max_flow(1,n);
    getch();
    return 0;
}

void short_path(int t,int l)
{
int i,u,v;
static int path[50];
v=t;
u=pred[v];
path[l++]=u;
if(u!=START)
 short_path(u,l);
else
{
 for(i=l-1;i>=0;i--)
  out<<path[i]<<"->";
  out<<n;
   }
  }
}
```

Example 6: Consider the graph as shown in Fig. 10.5. In the above program, the input stream is used to read data from input file "MF.TXT" and output stream is used to write data to a output file "MF_OUT.TXT". The first line of the input file "MF.TXT" is the number of vertices and edges. The rest is the adjacency matrix or capacity matrix of the graph shown in Fig. 10.5.

MF.TXT

```
6 10
1 2 16
1 3 13
3 2 10
2 3 4
4 3 9
2 4 12
3 5 14
5 4 7
4 6 20
5 6 4
```

After the execution of the program "MAXFLOW.CPP", the following output file "MF_OUT.TXT" is produced. The following output shows the Maximal Flow Pattern as well as the Maximal Flow from source vertex 1 to sink vertex 6.

MF_OUT.TXT

The number of nodes: 6

The number of edges: 10

The Adjacency matrix(Capacity matrix):
0 16 13 0 0 0
0 0 4 12 0 0
0 10 0 0 14 0
0 0 9 0 0 20
0 0 0 7 0 4
0 0 0 0 0 0

The Maximal Flow Pattern is the collection of all Augmenting Paths:

1.The Augmenting Path: 1->2->4->6

  The Residual Capacity of the Augmenting Path: 12

2.The Augmenting Path: 1->3->5->6

  The Residual Capacity of the Augmenting Path: 4

3.The Augmenting Path: 1->3->5->4->6

  The Residual Capacity of the Augmenting Path: 7

The Maximal flow: 23

References

Appel K, Haken W. Every planar map is four colorable. Bull Amer Math Soc. 1976;82:711–2.

Appel K, Haken W. Every planar map is four colorable. Part I. Discharging. Ill J Math. 1977a;21:429–90.

Appel K, Haken W. Every planar map is four colorable. Part II. Reducibility. Ill J Math. 1977b;21:491–567.

Balakrishnan R, Ranganathan K. A textbook of graph theory. New York: Springer; 2000.

Clark J, Holton DA. A first look at graph theory. Singapore: World Scientific; 1991.

Corman TH, Leiserson CE, Rivest RL, Stein C. Introduction to algorithms. 2nd ed. Cambridge: The MIT Press; 2001.

Deo N. Graph theory with applications to engineering and computer science. New Delhi: PHI; 1974.

Dharwadkar A. A new proof of the four color theorem. 2002 http://www.geocities.com/(2002).

Dharwadkar A, Pirzada S. Graph theory. India: Orient Longman and Universities Press of India; 2008.

Dharwadkar A, Pirzada S. Applications of graph theory. J Korean Soc Ind Appl Math (Ksiam). 2007;11(4): .

Dijkstra EW. A note on two problems in connexion with graphs. Numer Math. 1959;1:269–71.

Diestel R. Graph theory. In: Graduate texts in mathematics, vol. 173. New York: Springer; 1997 (2nd ed. 2000).

Dirac GA, Schuster S. A theorem of Kuratowski. Nederl Akad Wetensch Proc Ser A. 1954;57:343–8.

Euler L. Solutio problematics ad geometriam situs pertinents. Comment Academiae Sci 1 Petropolitanae. 1736;8:128–140.

Euler L. The Konigsberg bridges. Sci Amer. 1853;189:66–70.

Floyd RW. Algorithm 97: shortest path. Comm ACM. 1962;5:345.

Ford LR, Fulkerson DR. Flows in networks. Princeton: Princeton University Press; 1962.

Foulds LR. Graph theory applications. New York: Springer; 1992.

Harary F. A characterization of block graphs. Can Math Bull. 1931;6:1–6.

Harary F. Graph theory. Reading: Addison-Wesley; 1969.

Harary F, Frisch LT. Communication, transmission and transportation networks. Reading: Addison-Wesley; 1971.

Harary F, Hedetniemi S. The achromatic number of a graph. J Combin Theory B 1970; pp. 154–161.

Harary F, Hedetniemi S, Robinson RW. Uniquely colorable graphs. J Combin Theory. 1969;6:264–70.

S. Saha Ray, *Graph Theory with Algorithms and Its Applications*,
DOI: 10.1007/978-81-322-0750-4, © Springer India 2013

Harary F, Nash-Williams C St JA. On Eulerian and Hamiltonian graphs and line graphs. Can Math Bull. 1965;8:701–10.

Heawood PJ. Map color theorem. Qurat J Pure Appl Math. 1890;24:332–8.

Hopcroft JE, Tarjan RE. Efficient algorithms for graph manipulation. Comm ACM. 1973;16(6):372–8.

Kuratowski C. Sur le probleme des courbes gauches en topologie. Fund Math. 1930;15:271–83.

Moore EF. The shortest path through a maze. In: Proceedings of the international symposium on the theory of switching. Cambridge: Harvard University Press; 1959. pp. 285–292.

Map of USA with state names.svg—Wikipedia, the free encyclopedia http://en.wikipedia.org/wiki/File:Map_of_USA_with_state_names.svg.

Ore O. Note on Hamilton circuits. Amer Math Monthly. 1960;67:55.

Ore O. The four color problem. New York: Academic; 1967.

Ore O. Theory of graphs. Amer Math Soc Colloq Publ Providence. 1962;38.

Parthasarathy KR. Basic graph theory. New Delhi: Tata McGraw-Hill; 1994.

Saaty TL, Kainen PC. The four color problem. New York: McGraw-Hill; 1977.

Tanenbaum AS. Computer networks. Upper Saddle River: Prentice Hall Inc.; 1988.

Taha HA. Operations research: an introduction. Upper Saddle River: Prentice Hall Inc.; 2003.

Trudeau RJ. Introduction to graph theory. New York: Dover Publications; 1993.

Thomassen C. Kuratowskis theorem. J Graph Theory. 1981;5:225–41.

Warshall S. A theorem on Boolean matrices. J ACM. 1962;9:11–2.

West DB. Introduction to graph theory. Englewood Cliffs: Prentice Hall Inc; 1996.

Whitney H. A theorem on graphs. Annals Math. 1931;32:378–90.

Whitney H. Congruent graphs and the connectivity of graphs. Amer J Math. 1932;54:150–68.

Whitney VKM. Algorithm 422: minimal spanning tree. Comm ACM. 1972;15(4):273.

Index

Printed by Printforce, the Netherlands